EXPERIMENTAL ROCK MECHANICS

BALKEMA – Proceedings and Monographs
in Engineering, Water and Earth Sciences

Experimental Rock Mechanics

Kiyoo Mogi
Emeritus Professor of the University of Tokyo

CRC Press
Taylor & Francis Group
Boca Raton London New York

CRC Press is an imprint of the
Taylor & Francis Group, an **informa** business
A TAYLOR & FRANCIS BOOK

CRC Press
Taylor & Francis Group
6000 Broken Sound Parkway NW, Suite 300
Boca Raton, FL 33487-2742

First issued in paperback 2019

© 2007 by Taylor & Francis Group, LLC
CRC Press is an imprint of Taylor & Francis Group, an Informa business

Typeset by Charon Tec Ltd (A Macmillan Company), Chennai, India

No claim to original U.S. Government works

ISBN-13: 978-0-415-39443-7 (hbk)
ISBN-13: 978-0-367-39000-6 (pbk)
ISSN: 0929-4856

British Library Cataloguing in Publication Data
A catalogue record for this book is available from the British Library

Library of Congress Cataloging-in-Publication Data
Mogi, Kiyoo, 1929–
Experimental rock mechanics/Kiyoo Mogi.
p. cm.

Includes bibliographical references.
ISBN 0–415–39443–0 (hardcover : alk. paper) 1. Rock mechanics. I. Title.

TA706.M64 2006
624.1'5132—dc22

2006004322

Visit the Taylor & Francis Web site at
http://www.taylorandfrancis.com

and the CRC Press Web site at
http://www.crcpress.com

Table of contents

Preface   XI

About the author   XIII

PART I   DEFORMATION AND FRACTURE OF ROCKS

Chapter 1   Precise measurements of fracture strength
of rocks under uniform compressive stress   3

    1.1   Present specimen design   4
    1.2   Effect of length/diameter ratio to apparent strength
and fracture angle   6
    1.3   Comparison with the conventional method   11
    1.4   Decrease of the end effects by confining pressure   12
    References   15

Chapter 2   Deformation and failure of rocks under
confining pressure   17

    2.1   Deformation characteristics   17
       2.1.a   *Experimental procedure*   17
       2.1.b   *Stress-strain relation*   19
       2.1.c   *Modulus of elasticity*   23
       2.1.d   *Permanent strain*   26
       2.1.e   *Effects of previous loading*   26
          2.1.e.1   *Hydrostatic pressure*   27
          2.1.e.2   *Axial compression*   28
       2.1.f   *Yield stress*   29
       2.1.g   *Summary of the deformation characteristics*   31
    2.2   Pressure dependence of compressive strength and
brittle-ductile transition   32
       2.2.a   *Relation between strength and confining
pressure*   32
       2.2.b   *The Coulomb-Mohr fracture criterion*   37
       2.2.c   *Brittle-ductile transition*   43
    References   48

Chapter 3   Deformation and fracture of rocks under
            the triaxial compression: The effect of
            the intermediate principal stress   51

      3.1   History of compression experiments   51
            3.1.a   *Axial loading test under lateral pressure*   52
            3.1.b   *True triaxial compression test*   52
      3.2   Comparison between compression and extension under
            confining pressure   56
            3.2.a   *Introduction*   56
            3.2.b   *Experimental procedure*   57
                    3.2.b.1   *Confined compression test*   57
                    3.2.b.2   *Confined extension test*   58
            3.2.c   *Specimen materials*   60
            3.2.d   *Experimental results*   60
                    3.2.d.1   *Examination of isotropy and homogeneity*
                              by uniaxial compression tests   60
                    3.2.d.2   *Comparison of confined compression*
                              and extension tests   62
      3.3   True triaxial compression experiments   66
            3.3.a   *Introduction*   66
            3.3.b   *Design of the true triaxial apparatus*   67
            3.3.c   *Specimen design and strain measurement*   72
            3.3.d   *Experimental procedure and rocks studied*   74
            3.3.e   *Experimental results (1) – Stress-strain curves*
                    and fracture stresses   75
                    3.3.e.1   *Dunham dolomite*   76
                    3.3.e.2   *Solnhofen limestone*   82
                    3.3.e.3   *Yamaguchi marble*   88
                    3.3.e.4   *Mizuho trachyte*   94
                    3.3.e.5   *Manazuru andesite*   96
                    3.3.e.6   *Inada granite*   98
                    3.3.e.7   *Orikabe monzonite*   98
                    3.3.e.8   *Summary*   99
            3.3.f   *Experimental results (2) – Yield stresses*   103
                    3.3.f.1   *Dunham dolomite*   106
                    3.3.f.2   *Solnhofen limestone*   108
                    3.3.f.3   *Yamaguchi marble*   111
            3.3.g   *Failure criteria of rocks*   113
                    3.3.g.1   *Previous studies*   113
                    3.3.g.2   *Fracture criterion*   122
                    3.3.g.3   *Yield criterion*   128
                    3.3.g.4   *Summary*   130

3.3.h	*Ductility, fracture pattern and dilatancy*	131
	3.3.h.1 *Ductility and stress drop*	131
	3.3.h.2 *Fracture pattern*	135
	3.3.h.3 *Dilatancy*	151
3.3.i	*Fracture of an inhomogeneous rock and an anisotropic rock*	161
	3.3.i.1 *Inhomogeneous rock*	161
	3.3.i.2 *Anisotropic rock*	165
3.3.j	*Other recent experiments*	181
3.3.k	*Future problems*	185
References		186
Appendix		190

PART II ACOUSTIC EMISSION (AE)

Chapter 4   AE Activity   197

4.1	Introduction	197
4.2	AE activity under some simple loadings	198
4.3	Three patterns of AE activity	205
	References	215

Chapter 5   Source location of AE   217

5.1	Introduction	217
5.2	Experimental procedure	218
	5.2.a *Measurement of very high frequency elastic waves*	218
	5.2.b *Determination of source location of AE events*	220
	5.2.b.1 *One-dimensional case*	220
	5.2.b.2 *Two-dimensional case*	221
5.3	Experimental results	222
	5.3.a *Granite (heterogeneous silicate rock)*	222
	5.3.b *Andesite (moderately heterogeneous silicate rock)*	231
	5.3.c *Mizuho trachyte (nearly homogeneous silicate rock)*	232
	5.3.d *Yamaguchi marble with different grain sizes*	233
	5.3.e *Fracture of a semi-infinite body by an inner pressure source*	234
References		236

Chapter 6   Magnitude–frequency relation of AE events   239

    6.1   Introduction   239
    6.2   Experimental procedure and specimen materials   239
    6.3   The *m* value in the Ishimoto-Iida Equation   244
    6.4   Types of the magnitude–frequency relations and the
        structure of the medium   248
    6.5   Effects of measurements by different frequencies and
        different dynamic ranges of acoustic waves   250
        References   261

Chapter 7   AE Activity under cyclic loading   263

    7.1   Effect of tidal loading   263
        7.1.a   *Introduction*   263
        7.1.b   *Observations of AE events directly above the*
               focal region of the 1980 earthquake swarm   263
        7.1.c   *Seismic activity and ocean tide*   271
    7.2   AE under cyclic compression   277
        7.2.a   *Experimental procedure*   277
        7.2.b   *AE events under cyclic compression*   280
    7.3   AE under cyclic bending   286
        7.3.a   *Introduction*   286
        7.3.b   *Experiment A*   286
        7.3.c   *Experiment B*   288
        7.3.d   *Concluding remarks*   299
        References   302

Part III   ROCK FRICTION AND EARTHQUAKES

Chapter 8   Laboratory experiment of rock friction   307

    8.1   Introduction   307
    8.2   New design of a double-shear type apparatus   308
    8.3   Experimental result   315
        References   320

Chapter 9   Typical stick-slip events in nature and
            earthquakes   321

    9.1   Introduction   321
    9.2   Usu volcano and Unzen volcano, Japan   327

9.3   Sanriku-oki and Tokai-Nankai regions, Japan   329
9.4   Stick-slip and fracture as an earthquake mechanism   332
       References   333

Chapter 10   Some features in the occurrence of recent large
              earthquakes   335

       10.1 Global pattern of seismic activity   335
       10.2 Active and quiet periods in the main seismic zones   339
              10.2.a *Alaska – Aleutian – Kamchatka – N. Japan*   339
              10.2.b *Alps – Himalaya – Sunda*   341
       10.3 Some precursory seismic activity of recent large shallow
              earthquakes   343
              10.3.a *Introduction*   343
              10.3.b *2001 Bhuj (India) earthquake*   344
              10.3.c *2003 Tokachi-oki (Japan) earthquake*   346
              References   357

Subject index   359

Preface

This book does not aim to cover every problem of rock mechanics in a systematic manner. Rather, it touches upon selected subjects of a fundamental character and, in particular those pertinent to tectonophysics and seismology. Furthermore, it does not attempt to review all the developments but concentrates on my own contribution to the field of experimental rock mechanics. As a geophysicist, and a seismologist in particular, I have had an interest in the fracture and flow of rocks under stresses since the very beginning of my scientific career. In the field of seismology, I have written a book "Earthquake Prediction" (Academic Press, Tokyo, 1985). In parallel to these researches, I have pursuied experimental studies of the mechanical behaviour of rocks in the laboratory since about 1960.

In this book I will present, in detail, experimental methods developed and results obtained over the years by myself and my co-workers mainly at the Earthquake Research Institute of the University of Tokyo. In particular, many pages will be devoted to the problem of the deformation and fracture of rock specimens under the general triaxial compression in which all three principal stresses are different. Although this is one of the most fundamental and important problems, there has been no book so far in which it was fully discussed on the basis of reliable experimental data. In Chapter 3, the experimental results of the effect of the intermediate principal stress, which I obtained in the 1970s, are presented in a detailed manner, both in graphic and numerical forms. These results show the importance of the noticeable effect of the intermediate principal stress on the ultimate strength of rocks, that has been unjustifiably neglected or disregarded by Coulomb (1773), Mohr, Griffith and other strength theories which have been widely applied for a long time. They also show that the intermediate principal stress strongly influences the deformation and failure mode of rocks. On these experimental results, a new failure (fracture and yield) criterion is proposed. Furthermore, acoustic emission phenomena in rock under various stress states, and friction in rocks measured by a newly designed shear testing machine are discussed in relation to earthquake phenomena. Following on from these subjects, some noticeable features in the occurrence of recent large earthquakes are discussed.

Rock mechanics constitutes a branch of the material sciences. In the material sciences, the manufacturing of materials can be controlled in many cases. However, rocks are natural substances whose processes of generation and history of alteration in the Earth's crust are complex and generally unknown. In the early stages of the development of rock mechanics, numerous measurements in the laboratory were carried out without careful consideration of the complexity of rock materials. Therefore, for example, the compressive strength values of rock specimens taken from rock masses fluctuate greatly. Such data may be useful for practical purposes. However, empirical formulae derived from them should be discriminated from natural formulae that have a certain physical meaning. In order to promote rock mechanics as a modern material science, precise and specific experimental approaches should be especially devised. When investigating the fundamental properties of rocks, it is of utmost importance to reduce the extrinsic fluctuation of data and to increase the reproducibility of experimental results. For this purpose, the selection of suitable uniform rock samples and the development of new devices for reliable experimental tests with high accuracy are essential.

On the other hand, many rock materials and rock masses have inhomogeneous and anisotropic structures. In this book some experimental results are presented that show how these intrinsic features influence and may control the deformation mode and failure of rocks.

As mentioned above, this volume deals only with selected topics from the field of experimental rock mechanics. Consequently, the references are also limited; Drs M. S. Paterson and T.-f. Wong systematically reviewed almost all the remaining branches of rock mechanics, and listed many references in their book (M. S. Paterson and T.-f. Wong: Experimental Rock Deformation – The Brittle Field, Springer, Berlin 2005). Their book is suitable for anyone interested in subjects that are not covered in my book.

I am very grateful to Professor Marek Kwaśniewski, Editor-in-Chief of this Geomechanics Research Series, for inviting me to write this book and for his great help in the preparation of the manuscript, and to his assistant Mr. Ireneusz Szutkowski for his contribution of redrawing and/or remastering figures for the book. I am also grateful to Professor James J. Mori, Kyoto University, who read the latter part of the manuscript and gave valuable comments including English corrections. Technical support for experimental works by Hiromine Mochizuki and Kansaku Igarashi in the Earthquake Research Institute, Tokyo University is greatly appreciated. Finally, special thanks go to my wife Tomoko for her tireless support during the long writing and editing process.

About the author

Intrigued by the postulation that earthquakes occur by sudden fracture in the earth's crust, Kiyoo Mogi devoted an important part of his professional career to Experimental Seismology. He has carried out laboratory studies of the fracture phenomena of rocks under stress to establish their relation to natural earthquakes. Advocating that important features of fractures and earthquakes depend largely on the degree of mechanical heterogeneity of rock, Professor Mogi has investigated various features of seismic activity, such as seismic gaps and migration phenomena. His development of the true triaxial compression machine, gave greater insight into the mechanical behaviour of the earth's materials by the use of systematic experiments. Using the data obtained from triaxial compression tests, a new, general failure criterion was proposed, also known as the "Mogi Failure Criterion", which could be referred to generation and development of earthquakes. The current book represents the results of 30 years of investigation by the author and his team of the mechanical behaviour of rocks. Besides this work and many prominent papers in international journals, Kiyoo Mogi also authored the book "Earthquake Prediction", 1985, published by Academic Press, Tokyo.

Kiyoo Mogi (1929, Yamagata, Japan) studied Geophysics at the University of Tokyo. After obtaining his Ph.D. degree in 1962, he was employed as a research associate and an associate professor at the Earthquake Research Institute of the University of Tokyo, before becoming a Professor of Experimental Seismology from 1969 to 1990. After this period, he was Professor in Earth Sciences at the Nihon University in Narashino City, near Tokyo, until the year 2000. Kiyoo Mogi has held the positions of President of the Seismological Society of Japan and Director of the Earthquake Research Institute of University of Tokyo. He has also been the Chairman of the Coordinating Committee for Earthquake Prediction, Japan (1991–2000) and of the Earthquake Assessment Committee for the "Tokai Earthquake", Japan (1991–1996). Kiyoo Mogi is currently Professor Emeritus at the University of Tokyo, and Fellow of American Geophysical Union.

SELECTED PUBLICATIONS

Mogi, K. (1956). Experimental study of diffraction of water surface waves. *Bull Earthq. Res. Inst., Univ. Tokyo*, **34**, 267–277.

Mogi, K. (1958). Relations between the eruptions of various volcanoes and the deformations of the ground surfaces around them. *Bull. Earthq. Res. Inst., Univ. Tokyo*, **36**, 99–134.

Mogi, K. (1962). Study of elastic shocks caused by the fracture of heterogeneous materials and its relation to earthquake phenomena. *Bull. Earthq. Res. Inst., Univ. Tokyo*, **40**, 125–173.

Mogi, K. (1963). Some discussions of aftershocks, foreshocks and earthquake swarms-the fracture of a semi-infinite body caused by inner stress origin and its relation to the earthquake phenomena (3). *Bull. Earthq. Res. Inst., Univ. Tokyo*, **41**, 615–658.

Mogi, K. (1967). Earthquakes and fractures. *Tectonophysics*, **5**, 35–55.

Mogi, K. (1968). Sequential occurrences of recent great earthquakes. *J. Phys. Earth*, **16**, 30–36.

Mogi, K. (1968). Source locations of elastic shocks in the fracturing process in rocks (1). *Bull. Earthq. Res. Inst., Univ. Tokyo*, **46**, 1103–1125.

Mogi, K. (1970). Recent horizontal deformation of the earth's crust and tectonic activity in Japan (1). *Bull. Earthq. Res. Inst., Univ. Tokyo*, **48**, 413–430.

Mogi, K. (1971). Fracture and flow of rocks under high triaxial compression. *J. Geophys. Res.* **76**, 1255–1269.

Mogi, K. (1972). Fracture and flow of rocks. In *The Upper Mantle*, A. R. Ritsema (ed.), *Tectonophysics*, **13**, 541–568.

Mogi, K. (1974). On the pressure dependence of strength of rocks and the Coulomb fracture criterion. *Tectonophysics*, **21**, 273–285.

Mogi, K. (1977). Dilatancy of rocks under general triaxial stress states with special reference to earthquake precursors. *J. Phys. Earth*, **25**, Suppl., S 203–S 217.

Mogi, K. (1981). Seismicity in western Japan and long term earthquake forecasting. In *Earthquake Prediction, an International Review. M. Ewing Ser. 4*, D. Simpson and P. Richards (ed.), Washington, D.C., Am. Geophys. Union, 43–51.

Mogi, K. and H. Mochizuki (1983). Observation of high frequency seismic waves by a hydrophone directly above the focal region of the 1980 earthquake (M6.7) off the east coast of the Izu Peninsula, Japan. *Earthquake Predict. Res.*, **2**, 127–148.

Mogi, K. (1985). Temporal variation of crustal deformation during the days preceding a thrust-type great earthquake — The 1944 Tonankai earthquake of magnitude 8.1, Japan. *Pageoph*, **122**, 765–780.

Mogi, K. (1985). *Earthquake Prediction*. Academic Press, Tokyo, 355pp.

Kwasniewski, M. and K. Mogi (1990). Effect of the intermediate principal stress on the failure of a foliated anisotropic rock. In *Mechanics of Jointed and Faulted Rock*, H-P. Rossmanith (ed.), Balkema, Rotterdam, 407–416.

Mogi, K. (2004). Deep seismic activities preceding the three large "shallow" earthquakes off south-east Hokkaido, Japan — the 2003 Tokachi-oki earthquake, the 1993 Kushiro-oki earthquake and the 1952 Tokachi-oki earthquake. *Earth Planets Space*, **56**, 353–357.

Mogi, K. (2004). Two grave issues concerning the expected Tokai Earthquake. *Earth Planets Space*, **56**, Ii–Ixvi.

I

Deformation and Fracture of Rocks

Precise measurements of fracture strength of rocks under uniform compressive stress

The conventional compression test, in which a short, right cylinder is loaded axially, is one of the most widespread experimental procedures in rock mechanics. This configuration is used in studies of both brittle and ductile behavior, in long term creep studies as well as in studies of elastic behavior. In view of this wide usage, it is rather surprising to find so little concern with what are widely recognized as bad features of this test (Seldenrath and Gramberg, 1958; Fairhurst, 1961).

In the typical arrangement, for example, the steel of the testing machine contacts the rock cylinder, as shown in Fig. 1.1. Because steel and rock have different elastic properties, radial shearing forces are generated at the interface when load is applied to the rock sample. For most rocks these act inward and produce a *clamping effect* at the end of the cylinder (Filon, 1902). This causes two things to happen. First, because of the abrupt way in which the shearing stress changes near the outer edge of the steel-rock interface, a stress concentration arises (Nadai, 1924). Second, if a fracture propagates into the region near the end of the sample, growth of the fracture may be

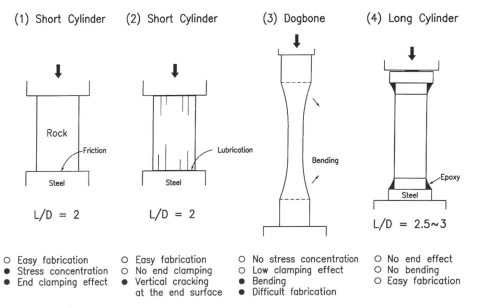

(1) Short Cylinder   (2) Short Cylinder   (3) Dogbone   (4) Long Cylinder

○ Easy fabrication	○ Easy fabrication	○ No stress concentration	○ No end effect
● Stress concentration	○ No end clamping	○ Low clamping effect	○ No bending
● End clamping effect	● Vertical cracking at the end surface	● Bending	○ Easy fabrication
		● Difficult fabrication	

Figure 1.1.   Various methods of testing rock samples under uniaxial compression conditions.

impeded. These two effects might influence the apparent strength of a typical rock differently: the stress concentration would tend to lower, while the clamping would tend to raise the apparent strength. Almost certainly, the effects do not cancel each other out.

One can think of a number of ways to eliminate these effects and thereby improve the compression test. As a simple method to eliminate the clamping end effect caused by friction on the steel-rock interface, various kinds of lubricants were applied between the rock end surface and the steel platen, as shown in Fig. 1.1 (2). In this case, a number of vertical cracks developed starting from the end surface of rock sample. This phenomenon seems to occur because of the intrusion of soft lubricator into the rock specimen. The compressive strength obtained by this method is different for different lubricants. Therefore, this method is not recommended.

Another simple method might be to make contact with the cylinder of rock through a metal which had identical mechanical properties. Then, no shearing forces would exist at the testing machine-rock interface, and, hence, there would be no stress concentration at the corners. However, faults might still be checked at the interface, for the typical short specimens that are used. The length/diameter ratio is usually around 2 so that any fault with a natural inclination of less that about 25° to the maximum compression would intercept one of the end faces. Actually, it is difficult if not impossible to match mechanical properties of rock with metals because most rocks exhibit complex non-elastic features prior to fracture. Therefore, this approach holds little promise.

Another approach is the improvement of sample shape, whereby the effect of a mismatch at the ends of the sample does not extend into the region of the sample where fracture occurs. One such design, suggested by Brace (1964), consists of a "dog-bone"-shaped specimen with a reduced central section and large radius as shown in Fig. 1.1 (3). According to three-dimensional photoelastic analysis (Hoek, personal communication to Brace), the stress concentration in the fillets of these specimens was extremely small and could be ignored. Of course, the stress concentration at the steel-rock interface and the associated clamping effect still existed in this specimen, but these had no effect on what happened in the central section, which, in Brace's design, had an area one fourth of that of the ends. Brace's specimens were quite slender so that bending stresses were present. Rather than go to extreme measures to eliminate them, they were simply evaluated by strain gages and the axial stresses corrected for bending in each experiments.

1.1   PRESENT SPECIMEN DESIGN

Brace's specimens have one drawback in that they require a grinding technique more elaborate than that used for producing straight cylinders. A new design was proposed by Mogi (1966) which combines the better features of Brace's specimen with greater ease of manufacture. The specimen (Fig. 1.2) is a long right cylinder connected to steel

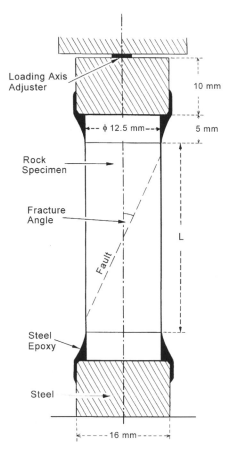

Figure 1.2.   The recommended test specimen design.

end pieces by epoxy which is a commonly available commercial variety which contains a filler of fine steel particles. The thickness of the epoxy is gradually decreased from the steel end pieces toward the middle of the specimen to form a smooth fillet.

The gradual decrease of the thickness of epoxy probably eliminates most of the stress concentration at the contact of rock with steel. As the steel-filled epoxy has a modulus somewhat less than most rocks, the exact surface shape of the fillet is not critical. The absence of a stress concentration near the end of the specimen is shown by the way in which samples fractured. Fractures remained near the central section and did not enter the fillet region of the sample. No partially formed fractures could be found in the fillet region. Yoshikawa and Mogi (1990) showed by a numerical calculation that there is no stress concentration near the fillet region and the stress distribution is homogeneous in the main part, as shown in Fig. 7.23.

Although the stress concentration near the ends is removed, the end part may give a clamping effect. To avoid the influence of clamping, the specimen has a length which allows propagation of fractures completely within a uniform stress field. This critical length is discussed in more detail below.

With the longer sample required to avoid the clamping effect, bending becomes likely, due mainly to a mismatch of loading and specimen axes. Bending was kept to a minimum here by (a) keeping the ends of the sample accurately parallel, (b) by applying the load at a small central area of the upper steel piece (called "loading axis adjuster" in Fig. 1.2), and (c) by keeping the length of the end pieces as small as possible. This is also discussed below.

In summary, this sample design appears to be more practical than Brace's specimen, and it eliminates most of the bad features present in the conventional short cylinder. Yet, it is almost as easy to fabricate as the conventional specimen, as the rock part is still a circular cylinder. Application of the method to extension tests will be mentioned in Section 3.2.b.

1.2   EFFECT OF LENGTH/DIAMETER RATIO ON APPARENT STRENGTH AND FRACTURE ANGLE

As mentioned above, apparent strength in short specimens becomes higher due to the clamping effect. With the increase of the length/diameter ratio, this effect should decrease gradually and disappear at some critical value. Above this critical value, the strength should remain constant and should represent the true strength under uniform compression. To obtain this value and the critical length/diameter ratio, relations between apparent strength and length/diameter ratio were experimentally investigated for uniaxial and triaxial compression. In these experiments, the triaxial testing apparatus designed by Brace was used for uniaxial and triaxial compression tests. Strain rate was held constant at about 10^{-4} sec^{-1}.

The author carried out careful measurements of the apparent compressive strength and fracture angles of Dunham dolomite, Westerly granite and Mizuho trachyte as functions of the length/diameter ratio using the above-mentioned test specimens (Mogi, 1966). Dunham dolomite and Westerly granite are compact and uniform in structure, while Mizuho trachyte is rather porous (porosity 8.5%) with a non-uniform distribution of pores. The results are summarized in Table 1.1 and Figs 1.3, 1.4 and 1.5. Here, L and D are length and diameter of the central cylindrical part of specimen, respectively.

Dunham dolomite (Fig. 1.3).   The average values of two or three strength measurements and the standard deviation are indicated in the figure. The reproducibility of strength (1 per cent or better) is very good. This is probably due to the homogeneous structure of this rock and the high accuracy of the present measurement. The apparent strength decreases markedly with the increase of L/D, but the value becomes nearly constant at high values of L/D. The critical value, (L/D)c, above which the apparent strength becomes nearly constant is about 2.5. In this rock, the fracture is of the typical shear type. The angle between the fracture and the loading axis also decreases markedly with increase of L/D and becomes nearly constant (20°) above (L/D)c. In Fig. 1.3 (lower), large closed circles indicate angles of fault planes which go through

Table 1.1. Apparent strength of specimens with different length/diameter ratio under uniaxial compression.

Rock	*L/D*	No. of Specimens	Apparent strength, MPa	Relative strength, %
Dunham dolomite	1.25	2	232 ± 1	111.5
	1.50	3	225 ± 3	107.5
	1.75	2	224 ± 3	107
	2.00	3	219 ± 3	104.5
	2.25	2	214 ± 0	102.5
	2.50	2	209 ± 0	100
	3.00	2	208 ± 0	99.5
	4.00	1	207	99
Westerly granite	1.25	2	263 ± 9	109.5
	1.50	2	252 ± 2	105
	1.75	2	248 ± 2	103.5
	2.00	3	247 ± 5	102.5
	2.25	3	242 ± 4	100.5
	2.50	4	240 ± 6	100
	3.00	2	239 ± 4	99.5
	3.50	2	238 ± 6	99
	4.00	3	238 ± 5	99
Mizuho trachyte	1.00	1	126	115.5
	1.50	1	114	104.5
	1.75	1	112	103
	2.00	1	110	101
	2.25	1	112	102.5
	2.50	1	110	100
	3.00	1	109	99.5

the whole sample and small closed circles indicate partial fault planes. In addition to actual faults, regular patterns of microfractures with no shear displacement were observed in the central part of certain specimens. The angle between these microfractures and the loading axis is indicated by open circles. Clearly, the orientation of the main faults depends on L/D, whereas the orientation of microfractures does not. Also, the former coincides with the latter for large L/D values. The dotted curve (1) in Fig. 1.3 (lower) gives a calculated value of θ from

$$\cot \theta = L/D \qquad (1.1)$$

where θ is an apparent fracture angle. Curve (2) is calculated from

$$\cot \theta = (L + 0.25)/D \qquad (1.2)$$

and takes into account the actual intersection that a straight fault would have some 3 mm above the base of the fillet, rather than exactly at the base of the fillet as in (1).

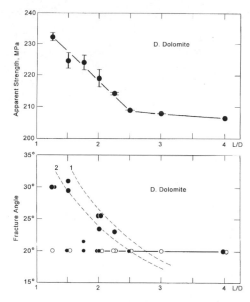

Figure 1.3.   Relations of apparent compressive strength (top figure) and apparent fracture angle (bottom figure) to length/diameter ratio in Dunham dolomite. In the bottom figure, large closed circle: good fault; small closed circle: partial fault; open circle: microfractures; dotted lines 1 and 2: calculated curves.

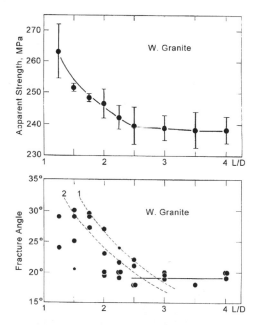

Figure 1.4.   Relations of apparent uniaxial compressive strength (top figure) and apparent fracture angle (bottom figure) to length/diameter ratio in Westerly granite.

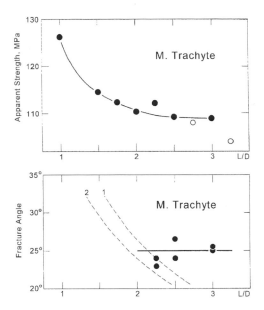

Figure 1.5.   Relation of apparent uniaxial compressive strength (top figure) and apparent fracture angle (bottom figure) to length/diameter ratio in Mizuho trachyte. Open circle: accompanied by bending.

The observed points are seen to lie somewhat between curves (1) and (2), which suggests that the fault angle in short specimens is strongly effected by clamping at the ends of the specimens. That is, the fault is forced to run between opposite corners of the sample.

Westerly granite (Fig. 1.4).   The effect of L/D on strength and fracture angle is very similar to that in dolomite, except for a greater fluctuation of strength values. This fluctuation (less than 3%) may be due to the larger grain size and high brittleness of this rock. The apparent strength becomes constant in longer specimens. The critical value of L/D is also 2.5. Apparent fracture angles also change with L/D and become constant (19°) in longer specimens. In this case, the typical fracture surface was not so flat as in the dolomite and no microfractures were observed.

Mizuho trachyte (Fig. 1.5).   The apparent strength also decreases with the increase of L/D and becomes constant above a critical value (2.0) of L/D, but this value is smaller than those obtained for the above-mentioned dolomite and granite. The final value of the fracture angle is larger (25°) than the other two. In this rock, a bending fracture was sometimes produced at higher values of L/D; such a fracture forms perpendicular to the specimen axis. The apparent strength of the other two was considerably lower than the normal case, as indicated by open circles in Fig. 1.5 (upper). The bending may have been caused by an uneven stress distribution in the specimen due to the porous structure.

Thus, the effect of L/D on apparent strength and fracture angle for these three rocks is important in shorter specimens. Above a critical value of L/D, strength and

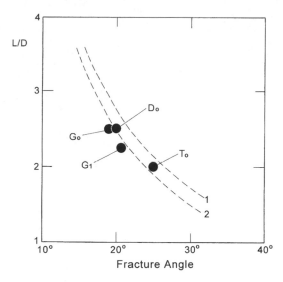

Figure 1.6.   Relation between fracture angle and the critical length/diameter ratio above which the end fix effect disappears. Broken lines 1 and 2: calculated curves; D_0: dolomite (confining pressure $= 0.1$ MPa); G_0: granite (0.1 MPa); G_1: granite (17 MPa); T_0: trachyte (0.1 MPa).

fracture angle were independent of L/D. This final value of strength and fracture angle probably represents the strength and fracture angle under uniform compressive stress for the following reasons: (1) The fracture starts and ends inside the central part of specimen. Therefore, stress concentrations at the end have no effects on strength. (2) Strength and fracture angle are independent of L/D, so the clamping effect is also avoided. (3) If bending is important, apparent strength should decrease with increase of the length of specimen. Here, since such a strength decrease is small in longer specimens (L/D < 4), so that bending effect is probably not significant.

The relation between the critical length/diameter ratio (L/D)c and the final fracture angle θ_0 is presented in Fig. 1.6. Curves (1) and (2) in this figure are calculated as given above. The agreement between the observation and the calculation is good, especially for curve (2).

A number of researchers have investigated the effect of length/diameter ratio on apparent uniaxial compressive strength for rocks (e.g., Obert et al., 1946; Dreyer et al., 1961) and for concrete (e.g., Gonnerman, 1925; Johnson, 1943; A.C.I., 1914). These previous experiments were carried out employing the conventional method in which straight circular cylinders or prisms were used, and therefore stress concentrations at the end of specimen were probably present. In Fig. 1.7, some results of these previous experiments (bottom figure) are presented together with the above-mentioned results (top figure). The apparent strength considerably decreases with the increase of length/diameter ratio (L/D). The strong effect of L/D in shorter specimens (L/D < 2) is clearly caused by the clamping effect. And these previous experiments did not give a constant final value of strength and fracture angle for large L/D. Apparent strength

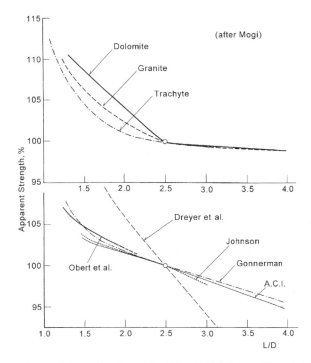

Figure 1.7.   Comparison of the author's results (Mogi, 1966) (top figure) with some previous experiments (bottom figure). The relative strength is recalculated for strength at $L/D = 2.5$. Obert et al. (1946): average for various rock types; Dreyer et al. (1961): rock salt; A.C.I. (1914), Gonnerman (1925), Johnson (1943): concrete.

still continues to decrease with L/D above the expected critical length/diameter ratio. This could have been due to bending caused by a mismatch of specimen and loading axis, or by the non-uniform structure of rock and concrete.

One example in the previous reports on similar measurements on an andesite by Shimomura and Takata (1961) is presented in the left figure of Fig. 1.8. The fluctuation of strength values ($\sim$50%) in their experiment is much greater than that of the above-mentioned Dunham dolomite shown in the right figure. From this experimental result, it is impossible to obtain any significant information. Such great fluctuation may be due mainly to high inhomogeneity of the rock samples. These previous experimental results presented in Fig. 1.7 and Fig. 1.8 show that a suitable selection of mechanically uniform rock samples is very important in rock mechanics.

1.3   COMPARISON WITH THE CONVENTIONAL METHOD

As mentioned above, the conventional test, which uses short cylinders, is accompanied by stress concentration and the clamping effect. Specimens of length/diameter ratio 2 were used in most previous experiments. According to the present results, such tests

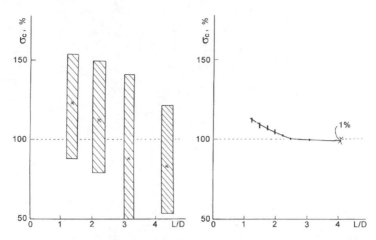

Figure 1.8. Simplified relations of apparent strength (relative values) to length/diameter ratio. Left figure: andesite (Shimomura and Takata, 1961); right figure: Dunham dolomite (Mogi, 1966).

may give considerably higher strength and larger fracture angle. Furthermore, since the end parts of the cylinder are laterally fixed to a hard end piece, the length of the central part where the stress is nearly uniform is still shorter than the actual length. For comparison with the present method, some uniaxial compression tests were carried out using the conventional method. In Dunham dolomite, the strength measured was 231 MPa and the fracture angle was 30°. These values are 10 per cent and 50 per cent higher, respectively, than those obtained using the present method. For Westerly granite, strength and fracture angle were 250 MPa and 26°. These are 5 per cent and 36 per cent higher than those obtained using the present method. Thus, applying the conventional test to such hard rocks may give considerably higher strength and larger fracture angle. In Mizuho trachyte, for which the true fracture angle is higher, the difference between the revised method and the conventional method was not appreciable.

Thus, the precise measurements of true compressive strength and fracture angle under uniaxial compression can be best carried out by use of the revised specimen design which is a longer cylinder with a smooth epoxy fillet at the end of rock specimen, as shown in Fig. 1.1 (4). And it should be noted that the length/diameter ratio should be larger than 2.5.

1.4 DECREASE OF THE END EFFECTS BY CONFINING PRESSURE

It was expected that end effects might change with confining pressure. To investigate this, apparent strength and fracture angle of the specimens having various length/diameter ratio were measured under confining pressure (Mogi, 1966). The shape of specimen and the experimental procedure are similar to the case of uniaxial

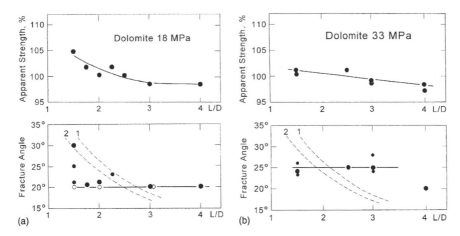

Figure 1.9. Relation of apparent strength (top figure) and apparent fracture angle (bottom figure) to length/diameter ratio in Dunham dolomite under confining pressure of 18 MPa (a) and 33 MPa (b).

compression, except for jacketing of the specimen to prevent the intrusion of pressed oil. Corrections were made for the frictional force between the axial piston and the o-ring seal of the pressure vessel. (The experiments under confining pressure are discussed in detail in the following chapter.)

Dunham dolomite. The results are presented in Figs. 1.9 (a) and 1.9 (b). Under 18 MPa confining pressure, the final fracture angle and the critical value of L/D are nearly the same as in the uniaxial case. However, the effect of L/D on strength is considerably smaller than in the uniaxial case. Strength and fracture angle under 33 MPa confining pressure is nearly independent of L/D.

Westerly granite. The effect of L/D was determined at a confining pressure of 17 MPa, 51 MPa, and 108 MPa. With the increase of pressure, the effect of L/D on strength and fracture angle decreases, as shown in Figs. 1.10 (a)–1.10 (c). The fracture angle gradually increases with confining pressure and the critical value of L/D also decreases.

Thus, it is concluded that end effects decrease gradually with an increase of pressure and become very small under confining pressure greater than 30–50 MPa, as shown in Fig. 1.11. This decrease of end effect may be due to (1) the relative decrease of the effect of lateral restriction at the end part by increase of lateral pressure, (2) the increase of fracture angle under pressure, and (3) the increase in the ductility of rocks. The observations show that the shape of the specimen becomes less critical at high pressure and so the conventional triaxial compression test may also give as nearly correct results as the method used here.

This discovery of the marked decrease of the end effect by confining pressure provided *a key for the design of the true triaxial compression machine* (Mogi, 1970), which is explained in Chapter 3.

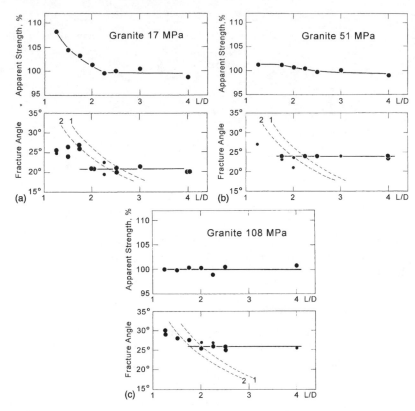

Figure 1.10.   Relation of apparent strength (top figure) and apparent fracture angle (bottom figure) to length/diameter ratio in Westerly granite under confining pressure of 17 MPa (a), 51 MPa (b) and 108 MPa (c).

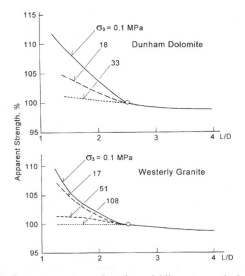

Figure 1.11.   Relation of apparent strength to length/diameter ratio in Dunham dolomite (top figure) and Westerly granite (bottom figure).

REFERENCES

Am. Concrete Inst. (A.C.I.) (1914). Report of committee on specifications and methods of tests for concrete materials. Proc. Am. Concrete Inst., v. **10**, p.422.

Brace, W. F. (1964). Brittle fracture of rocks. In: State of stress in the Earth's Crust. Judd, W.R. (ed.) New York: Elsevier, 111–174.

Dreyer, W. and H. Borchert. (1961). Zur Druckfestigkeit von Salzgesteinen. Kali und Steinsalz, Heft **7**, S.234–241.

Fairhurst, C. (1961). Laboratory measurement of some physical properties of rock. In 4th Symp. on Rock Mechanics, Bull. Mineral Industries Expt. Sta. Penn. State Univ., No. 76. Hartman, H. L. (ed.), 105–118.

Filon, L. N. G. (1902). On the elastic equilibrium of circular cylinders under certain practical systems of load. Phil. Trans. R. Soc., London, Ser. A **198**, 147–233.

Gonnerman, H. F. (1925). Effect of size and shape of test specimen on compressive strength of concrete. Proc. A. S. T. M., v. **25** (Part 2), p.237.

Johnson, J. W. (1943). Effect of height of test specimen on compressive strength of concrete. A. S. T. M. Bulletin, No. 120, p.19.

Mogi, K. (1966). Some precise measurements of fracture strength of rocks under uniform compressive stress. Felsmechanik und Ingenieurgeologie **4**, 41–55.

Nadai, A. (1924). Über die Gleitund Verzweigungsflächen einiger Gleichgewichtszustände, Z. Physik 30, p.106.

Obert, L., S.L. Windes and W. I. Duvall. (1946). Standardized tests for determining the physical properties of mines rocks. U. S. Bur. Mines. Rep. Invest., no. 3891, p.1.

Seldenrath, Th. R. and J. Gramberg. (1958). Stress-strain relations and breakage of rocks. In: Mechanical Properties of Non-Metallic Materials. Walton, W. H. (ed.). London: Butterworths, 79–102.

Shimomura, Y. and A. Takata. (1961). On the mechanical behaviors and the breaking mechanism of rocks (1st Report) — Shape and scale effect on fracture of rock specimens in compression. J. Mining and Metallurgical Inst. of Japan, Vol. **77**, No. 876, 9–14. (in Japanese)

Yoshikawa, S. and K. Mogi. (1990). Experimental studies on the effect of stress history of acoustic emission activity – A possibility for estimation of rock stress. J. Acoustic Emission, *8*, 113–123.

CHAPTER 2

Deformation and failure of rocks under confining pressure

A state of stress that is necessary to produce rock failure in an element can be described by the three principal stresses σ_1, σ_2, and σ_3. In this book, compressive stress is taken positive and $\sigma_1 > \sigma_2 > \sigma_3$, that is, σ_1 is the maximum principal stress, σ_2 is the intermediate principal stress, and σ_3 is the minimum principal stress. In principal coordinates, the points (σ_1, σ_2, σ_3) representing different states of stress necessary to produce failure might form a surface

$$\sigma_1 = f(\sigma_2, \sigma_3) \tag{2.1}$$

for a given material. The most fundamental problem of rock mechanics is the study of the shape of this surface for various rocks. In the next chapter, this is fully discussed.

Many experiments have been done using the so-called triaxial test, in which two principal stresses are equal, $\sigma_1 > \sigma_2 = \sigma_3 > 0$ (e.g., von Kármán, 1911; Griggs, 1936; Handin, 1966) or $\sigma_1 = \sigma_2 > \sigma_3 > 0$ (e.g., Böker, 1915). In this book, the term "conventional triaxial test" is used for such cases, and the true triaxial test is used for the general case in which $\sigma_1 \geq \sigma_2 \geq \sigma_3$. In this chapter, deformation characteristics and pressure dependence of rock strength by the conventional triaxial compression test are discussed, mainly on the basis of the author's experiments.

2.1   DEFORMATION CHARACTERISTICS

2.1.a   *Experimental procedure*

First, rock specimens of various types from Japan were tested at room temperature by the conventional triaxial compression method (Mogi, 1965). The testing apparatus at Waseda University in Tokyo was used in the experiments. The schematic view of the apparatus is presented in Fig. 2.1. The test specimens were cylinders of 40 mm diameter for weaker rocks and 20 mm diameter for hard rocks, the height/diameter ratio being nearly 2.0. In this case, there is the end effect for hard rocks under low confining pressure, but the revised specimen design described in the preceding chapter was used for Westerly granite and Dunham dolomite which will be discussed later in this chapter. The specimens were jacketed with soft rubber to prevent pressure fluid from entering porous rock specimens. (Thereafter, I frequently used silicone rubber for jacketing.) The axial stress was applied by a hydraulic testing machine. The confining pressure was supplied independently by another oil compressor. The value of confining pressure was measured by a Bourdon gage. The axial load was

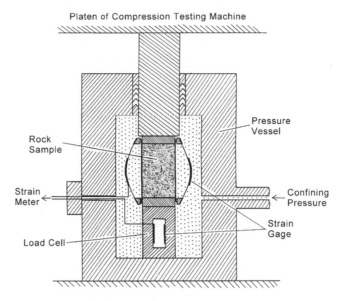

Figure 2.1.   Schematic view of the triaxial apparatus.

measured by a load cell situated between a steel end piece fixed to the test specimen and the bottom of the pressure vessel. The load cell, a closed hollow steel cylinder with an electric resistance strain gage bonded to the inside of the cylinder, measured the axial load without any frictional error.

In Fig. 2.2, various methods of the axial strain measurement are shown. The external method (1) in this figure, using a dial gage or differential transformer in which the strain is obtained from the displacement of the piston of the press, is not accurate enough for a precise analysis of the stress-strain curve and it is not applicable for cyclic loading because this system shows a remarkable hysteresis. For precise measurements of strain, the internal method (2) of which the two methods (a) and (b) are schematically shown in Fig. 2.2. In the first method (a), the electric resistance strain gage is bonded directly onto the surface of the rock specimen. This method is highly sensitive and easy-to-use, therefore it is used most often. However, method (a) is unsuitable to measure the mean strain in heterogeneous deformation including microcracks, small faults, etc. and to measure large deformation. In method (b), bending steel plates are fixed to the upper and lower steel end pieces connected to the rock specimen, and shortening of the rock cylinder causes the bending of the thin steel plates. The bending deformation of the plate is measured by the bonded strain gages. As the elastic distortion of steel end pieces is generally negligible because of its slight thickness, the strain gage output indicates the axial strain of the rock specimen. Since this system of strain measurement does not show any appreciable hysteresis for loading and unloading, this method is applicable for cyclic loading experiments, and also is suitable for large deformation measurements.

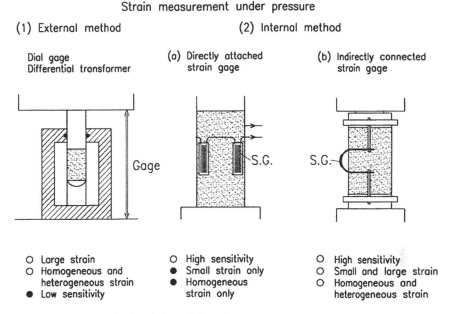

Figure 2.2.   Various methods of the axial strain measurement.

However, the behavior of electric resistance type strain gage should be examined under confining pressure, because there was a question that the strain gage might show an appreciable "pressure effect" due to hydrostatic pressure. To examine this "pressure effect", the change in strain measured by the electric resistance strain gage under confining pressure was compared with that in the atmospheric pressure, as shown in Fig. 2.3. Strains (A) and (B) are the axial strains of, respectively, the outside and the inside of the hollow steel cylinder with closed ends under axial compression. Open circles in Fig. 2.3 are in atmospheric pressure (0.1 MPa) and closed circles are in 130 MPa confining pressure. Since the relation between strain (A) and strain (B) is almost similar in both cases, the hydrostatic pressure does not give any significant effect to the gage factor.

Thus, method (b) shown in Fig. 2.2 or Fig. 2.1 is applicable to precise measurements of small and large strains of various rocks under cyclic loadings. The axial load was applied at a nearly constant strain rate of 0.15–0.2 percent per min. The differential stress was reduced to zero at various stages of deformation, the elastic and permanent strains were observed at each stage.

2.1.b   *Stress-strain relation*

The rocks tested were a peridotite, a diorite, a granite (strength only), two andesites, a trachyte, three tuffs and a marble. The density and porosity of these rocks are shown in Table 2.1.

The test specimens were compressed to fracture for hard rocks and to 3–4 percent strain for soft rocks, and at various stages of deformation the stress was reduced to

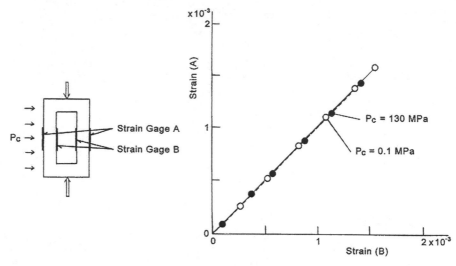

Figure 2.3.   Strains (A) and (B) are the axial strains at, respectively, the outer and inner surface of the steel hollow cylinder with closed ends subjected to confining pressure.

Table 2.1.   Tested rocks.

No.	Rock	Bulk density (g/cm^3)	Porosity (%)
1.	Nabe-ishi peridotite	3.16	0.02
2.	Orikabe diorite	2.78	0.4
3.	Mannari granite	2.62	0.7
4.	Mito marble	2.69	0.2
5.	Shirochoba andesite	2.45	5.1
6.	Tatsuyama tuff	2.26	10.2
7.	Mizuho trachyte	2.24	8.5
8.	Shinkomatsu andesite	2.17	12.6
9.	Ao-ishi tuff	2.01	17.3
10.	Saku-ishi welded tuff	1.95	21.6

zero or to very small values, then increased again. Stress-strain curves of some of these rocks under the conventional triaxial compression for different values of the confining pressure ($\sigma_2 = \sigma_3$) are shown in Figs. 2.4(a)–2.4(f). The vertical axis is the differential stress ($\sigma_1 - \sigma_3$) in MPa and the horizontal axis is the strain in percent.

The experimental results show that compact igneous rocks are very brittle and become markedly stronger with the increase of confining pressure, while other porous soft rocks become ductile even under low confining pressure. However, the precise feature of these curves suggests complex processes of deformation. For example, the initial part of deformation before yielding, the so-called *elastic part*, is significantly curved and includes appreciable permanent deformation in many cases. The simple

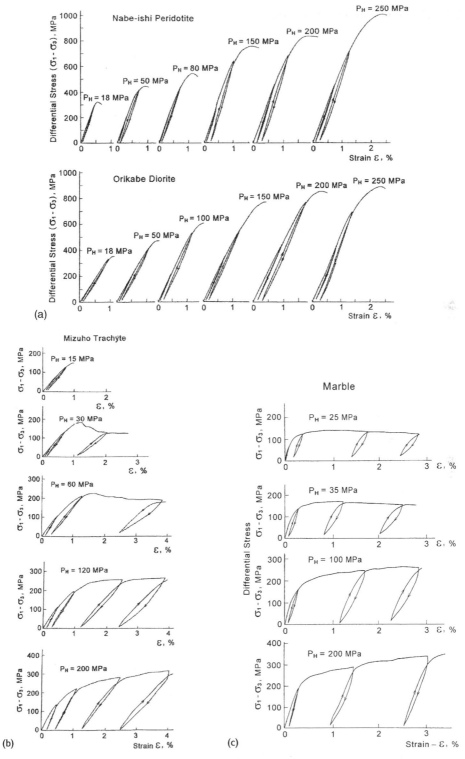

Figure 2.4(a)–(f).   Stress-strain curves under cyclic axial loading of various kinds of rocks.
P_H: confining pressure.

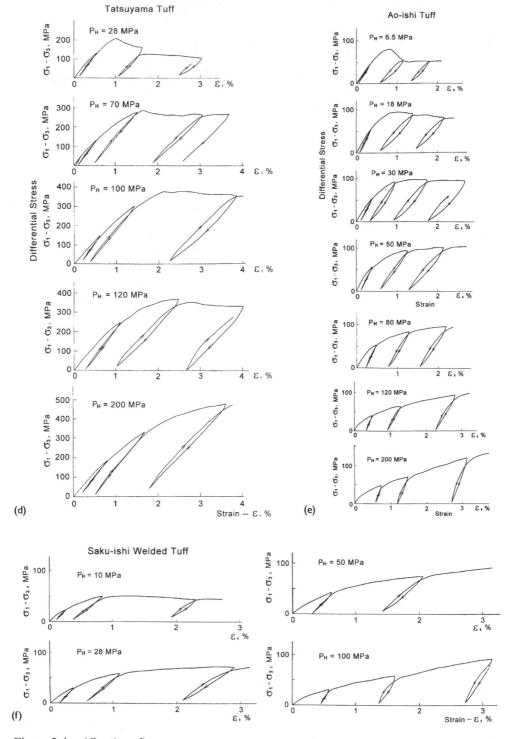

Figure 2.4.   (*Continued*)

slope of the stress-strain curve is different from the elastic modulus which is defined in the later discussion.

The terms *brittle* and *ductile* which are used in this book are understood as follows. Brittle behavior is characterized by a sudden change of slope in the stress-strain curve near the yield point followed by a complete loss of cohesion or an appreciable drop in differential stress. Ductile behavior is characterized by the deformation without any downward slope after the yield point. Further discussions of the definition of *brittle*, *ductile*, and *yielding* will be presented later.

2.1.c *Modulus of elasticity*

When the rock specimen is compressed to a point P in Fig. 2.5 and then unloaded to zero differential stress and then reloaded, the branches (P2Q, Q3P) are different from the virgin loading curve (O1P) in many cases, as shown schematically in Fig. 2.5. The unloading and reloading curves usually differ little from each other and form a narrow loop (P2Q3). The reloading curve (Q34) passes very near the point P from which the unloading branch drops down and continues to the virgin loading curve (O1P). This narrow loop circumscribed by the unloading and reloading branches may be approximately substituted by a straight line $\overline{PQ}$. The slope of this line is taken as the mean Young's modulus.

Figure 2.6 show the relation between Young's modulus (E) and the total strain (ε), which is defined in Fig. 2.5, for different values of confining pressure. According to this result, Young's modulus differs considerably at various stages of deformation. In compact igneous rocks, Young's modulus is nearly constant until their fracture. In other porous rocks, it generally decreases with increasing strain (ε), as seen in Fig. 2.6. However, the modulus (E) of some ductile rocks (tuffs and marble) under high confining pressure begins to increase with strain just after the yielding. The marked decrease of Young's modulus (E) with increasing strain (ε) is attributed to the microfracturing in these rock specimens, and the increase of Young's modulus (E) in a large

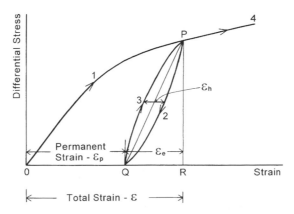

Figure 2.5.   Typical stress-strain curves under cyclic loading.

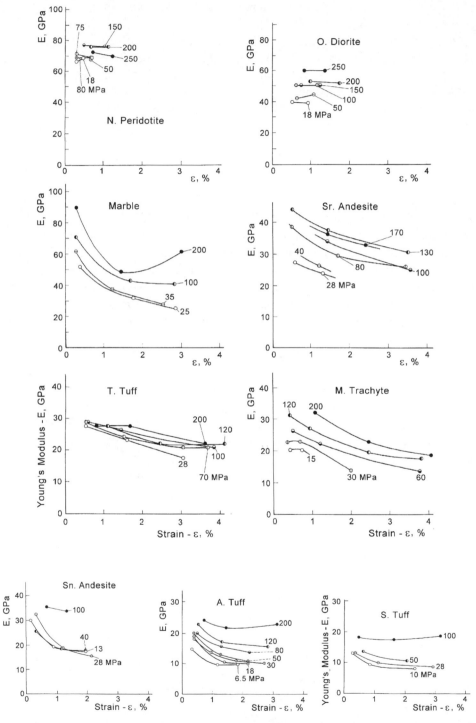

Figure 2.6.   Relation between Young's modulus (E) and the total strain (ε).

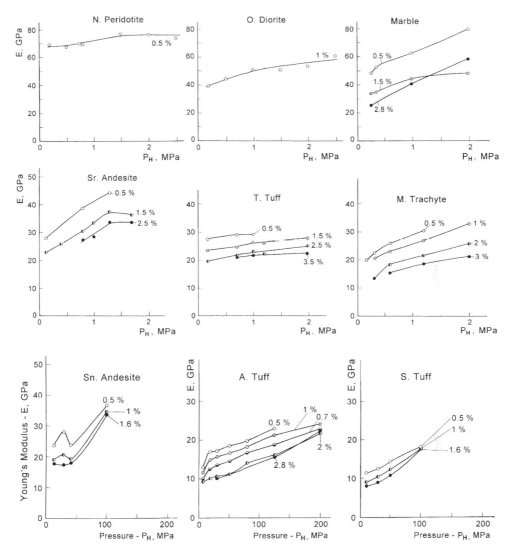

Figure 2.7.   Relation between Young's modulus and the confining pressure. Values of the total axial strain are given for each curve.

deformation stage under high confining pressure in soft rocks can be attributed to compaction of fractured rock specimens.

The relations between Young's modulus (E) and confining pressure ($\sigma_3 = \sigma_2$) are shown for different values of strain in Fig. 2.7. Normally, Young's modulus increases with pressure. The modulus of Nabe-ishi peridotite, which is the most compact rock (porosity 0.02%) among the tested rocks, is largest and increases slightly with confining pressure. The modulus E of Orikabe diorite (porosity 0.4%) is lower than that of peridotite and its increase with confining pressure is more appreciable. The results obtained for igneous compact rocks are consistent with similar preceding experiments (e.g., Brace, 1964), and with results of elastic wave velocity measure-ment (Birch, 1960). In more porous rocks, Young's modulus increases more markedly with confining pressure, as shown in Fig. 2.7.

The pressure dependence of Young's modulus in Shinkomatsu andesite is abnormal at lower pressures, as seen in Fig. 2.7. Young's modulus increases with a pressure increase to 28 MPa and drops noticeably at 40 MPa. This drop in Young's modulus is not so clear at larger strains. This abnormal change in Young's modulus seems to be attributed to the mechanical structure of this rock. That is, the porosity of this volcanic rock is markedly high (12.6%) and most pores are isolated round voids in a continuous framework, while other porous sedimentary rocks have wedge-like openings between grains. The above-mentioned abnormal behavior of the andesite seems to be due to the fracturing of this rock's framework by the confining pressure.

2.1.d *Permanent strain*

The strain of stressed rocks is partly elastic and partly permanent. The elastic strain is obtained as a recovered deformation by unloading to zero differential stress while the permanent strain is obtained as an unrecovered deformation (Fig. 2.5). This perma-nent strain includes various types of unrecoverable deformation due to dislocation, viscous flow, micro-fracturing, etc. and so it increases to some degree with the decrease of strain rate (e.g., Heard, 1963).

In Fig. 2.8, the relation between the permanent strain (ε_p) and the total strain (ε) is shown for different confining pressures. It should be noted that the permanent strain is found even at an initial stage of deformation in most rocks. In compact igneous rocks which are very brittle, the permanent strain appears slightly in the nearly linear part of stress-strain curve and increases abruptly from a proportional limit. In some porous silicate rocks, such as Tatsuyama tuff (porosity 10.2%), the appreciable permanent deformation increases linearly at the initial stage and the slope of this curve increases suddenly at some stage. In highly porous silicate rocks, such as Ao-ishi tuff (porosity 17.3%), the permanent strain is very large and increases continuously without any such sudden change in the slope of curve, particularly under higher confining pressure. In carbonate rocks, which show remarkable ductile behavior, the $\varepsilon_p - \varepsilon$ relation is very different from those of silicate rocks. The case of Mito marble (porosity 0.2%) is shown in Fig. 2.8. The ratio $\varepsilon_p/\varepsilon$ is very large (60–80%) except for the initial stage of deformation, and nearly independent of confining pressure. The main part of large deformation of marble is permanent, as can be seen in Figs. 2.4(c) and 2.8.

2.1.e *Effects of previous loading*

As mentioned above, rocks are not completely elastic materials and they show remark-able plasticity; during the deformation of rocks microfracturing occur in many cases. Therefore, the stress-strain relation is affected by a history of deformation. If the effect of the previous deformation is clarified, a history of deformation of rocks may be deduced from the present mechanical property. As a preliminary step from this

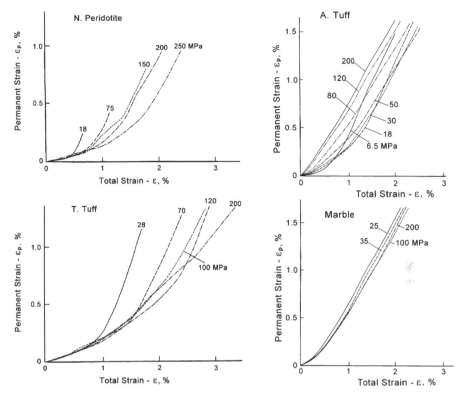

Figure 2.8.   Relation between the permanent strain and the total strain. Values of the confining pressure are given for each curve.

viewpoint, an effect of previous loading of Ao-ishi tuff was measured in some simple cases (Mogi, 1965).

2.1.e.1   *Hydrostatic pressure*

An effect of previously applied hydrostatic pressure was studied. Ao-ishi tuff specimens were initially exposed to hydrostatic pressure of 80 MPa and 200 MPa. Thereafter, these specimens and a virgin specimen were tested under conventional triaxial compression conditions at 18 MPa confining pressure. Curves (B) and (C) in Fig. 2.9 show the stress-strain characteristics of specimens which were initially exposed to 80 MPa and 200 MPa hydrostatic pressure respectively, and curve (A) is that of the virgin specimen. Both the compressive strength and Young's modulus in the case of 80 MPa hydrostatic pressure are not significantly different from those of the virgin rock. On the other hand, the compressive strength and Young's modulus in the case of 200 MPa are markedly lower than those of the virgin rock. From this result, it may be deduced that this rock has not been buried at a larger depth than several kilometers.

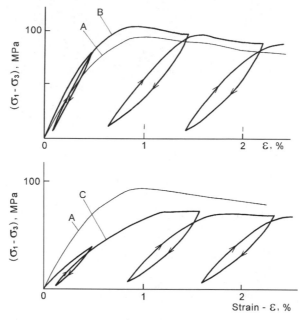

Figure 2.9.   Stress-strain curves of Ao-ishi tuff under 18 MPa confining pressure. (A): virgin specimen; (B): specimen previously exposed to 80 MPa hydrostatic pressure; (C): specimen previously exposed to 200 MPa hydrostatic pressure.

2.1.e.2   *Axial compression*

As can be seen from the stress-strain curves obtained by cyclic loading, Young's modulus and plasticity are affected by a previous compression. After initial compression, the stress-strain curve markedly changes in comparison to that of virgin rocks. The experimental result of Ao-ishi tuff is shown in Fig. 2.10. A specimen was axially shortened by 2.2 percent by the conventional triaxial compression test (confining pressure 80 MPa). After being left 18 hours at atmospheric pressure, the specimen was tested again by compression under 80 MPa confining pressure [curve (2)]. Young's modulus decreases somewhat due to previous loading, but its value corresponds to that in deformation continued without such a break in loading. The stress-strain curve before yielding in curve (2) is nearly linear and the permanent strain at this stage is smaller as compared with the initial part of the deformation of the virgin rock. The most noticeable change between the two curves (1) and (2) is the marked increase of the apparent yield stress in the previously loaded specimen. From this result, it may be deduced that the yield stress corresponds to the magnitude of previously applied stress in such cases.

Effects of previous loading may be complex. However, a history of deformation of rock specimen is more or less preserved in mechanical properties, such as elasticity, plasticity, fracture strength, deformation characteristics in repeated loadings, and other qualities. Thus, the deformation experiments from this point of view may be useful for studying the mechanical history of the earth's materials. After many years of

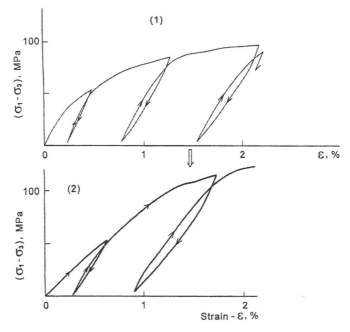

Figure 2.10. Stress-strain curves of Ao-ishi tuff under 80 MPa confining pressure. (1): virgin specimen; (2): specimen previously shortened by 2.2%.

this preliminary experiment, more careful experiments have been conducted from the similar standpoint. For example, Yamamoto et al. (1990) developed a method for estimating crustal stress by measurements of the mechanical history of rock specimens.

2.1.f *Yield stress*

Strength in the brittle state is the maximum stress achieved during an experiment. This value is definitely determined. However, failure strength in the ductile state, namely yield stress, is usually not so definite. Generally, the yield stress is the stress at which the sudden transition from elastic to plastic deformation state takes place, and so the stress at the knee of the stress-strain curves is taken as the yield stress (e.g., Robertson, 1955). In some rocks, the knee of the stress-strain curve is clear and the determination of yield stress is easily possible. However, other rocks show a gradual transition from elastic to plastic stage, that is, the slope of the stress-strain curve varies gradually. In this case, the definite determination of yield stress is difficult.

Sometimes the yield point is defined as a point having an appreciable limit value of permanent deformation. This definition is also not always suitable for various rocks, because the value of the permanent strain at the marked break of the stress-strain curve is not constant and varies widely.

If the yielding is understood in a wide sense, the yield stress of rocks may be more generally defined in the following way. The local yielding begins to occur even under

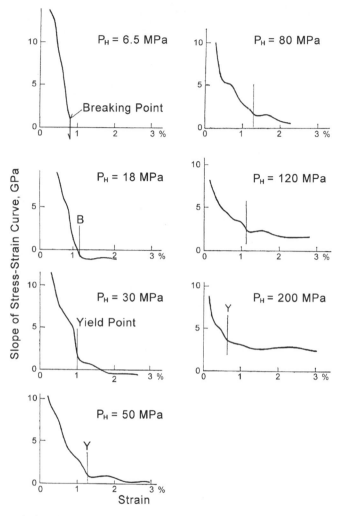

Figure 2.11.   Relation between $\Delta(\sigma_1 - \sigma_3)/\Delta\varepsilon$ and strain (ε) for Ao-ishi tuff. Valucs of confining pressure are given for each curve.

low stress, because the stress in the rock specimen having nonuniform structures distributes concentratively around structural irregular points. The yielding in such nonuniform materials takes place gradually in a wide range of stress values, and the macroscopic yielding may be completed at a certain stress value. This terminal stress is treated as yield stress. In fact, the slope of the stress-strain curve decreases gradually with the increase of strain and becomes nearly constant at the yield stress in many cases. This definition is also consistent with the usual definition in the case characterized by a marked break of the stress-strain curve. Fig. 2.11 shows the relation between the slope of the stress-strain curve [$\Delta(\sigma_1 - \sigma_3)/\Delta\varepsilon$] and the axial strain ($\varepsilon$) of Ao-ishi tuff. It is suitable that, the point at which $\Delta\sigma/\Delta\varepsilon$ becomes nearly constant is taken as the yield point.

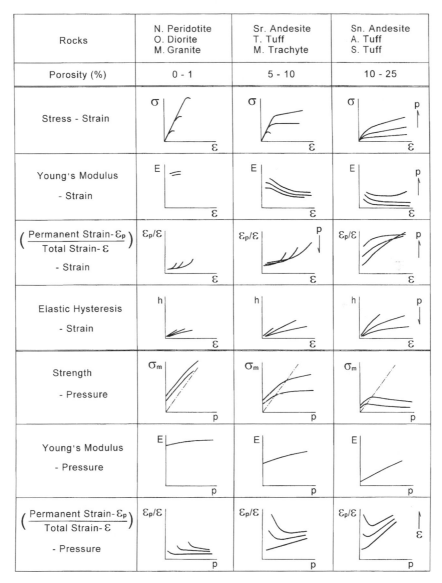

Rocks	N. Peridotite O. Diorite M. Granite	Sr. Andesite T. Tuff M. Trachyte	Sn. Andesite A. Tuff S. Tuff
Porosity (%)	0 - 1	5 - 10	10 - 25
Stress - Strain			
Young's Modulus - Strain			
$\left(\dfrac{\text{Permanent Strain-}\varepsilon_p}{\text{Total Strain-}\varepsilon}\right)$ - Strain			
Elastic Hysteresis - Strain			
Strength - Pressure			
Young's Modulus - Pressure			
$\left(\dfrac{\text{Permanent Strain-}\varepsilon_p}{\text{Total Strain-}\varepsilon}\right)$ - Pressure			

Figure 2.12.   Summary of the deformation characteristics of silicate rocks under confining pressure and at room temperature (Mogi, 1965).

2.1.g   *Summary of the deformation characteristics*

Using the conventional triaxial test and the indirect strain gage method, the stress-strain curves in cyclic loading were obtained for different types of rocks. The tested silicate rocks can be divided into three groups according to their mechanical properties. The mechanical properties of these rocks are roughly summarized in Fig. 2.12. The range of porosity in these rocks is tabulated in this figure. From this result, it can be clearly seen that porosity is the most important factor for mechanical properties in silicate rocks. The mechanical properties of the marble are significantly different from those of silicate rocks. Although the porosity of the marble is very small (about 0.2%), its properties are nearly similar to the third group in appearance. This shows that the mineral composition is also an important factor in mechanical behavior in rock deformation under pressure.

2.2   PRESSURE DEPENDENCE OF COMPRESSIVE STRENGTH AND BRITTLE-DUCTILE TRANSITION

2.2.a   *Relation between strength and confining pressure*

As mentioned in the preceding section, the compressive strength of rocks generally increases with increasing confining pressure. The pressure dependence of

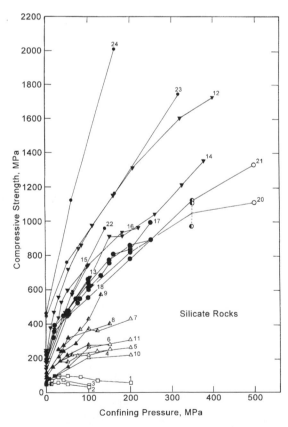

Figure 2.13.   Relation between compressive strength and confining pressure for silicate rocks. Numbers given for each curve correspond to rocks listed in Table 2.2. Closed symbols: brittle behavior; semi-closed: transitional behavior; open: ductile behavior.

rock strength has been experimentally studied by many investigators (e.g., Adams and Nicolson, 1901; von Kármán, 1936; Robertson, 1955; Griggs, 1936; Handin and Hager, 1957; Brace, 1964; Mogi, 1966a,b). Fracture strength of brittle hard rocks increases markedly with increasing pressure. On the other hand, the yield stress of ductile rocks, such as some limestones and marbles, does not increase appreciably.

Compressive strengths ($\sigma_1 - \sigma_3$) of silicate rocks and carbonate rocks are shown as functions of confining pressure ($\sigma_2 = \sigma_3$) in Fig. 2.13 and Fig. 2.14, respectively (Mogi, 1966a). Data in these figures are taken from references listed in Table 2.2. In the figures, the closed, semi-closed and open symbols indicate brittle, transitional and ductile behavior, respectively. These rocks are divided into several groups for convenience, as shown in Table 2.2. These strength-pressure curves are more or less concave downward with a roughly similar shape. Fig. 2.15 shows the relation between ($C-C_0$) and confining pressure of silicate rocks in logarithmic scales, where C and C_0 are compressive strength at elevated confining pressures and at atmospheric pressure, respectively. As shown for Kitashirakawa granite by Matsushima (1960), these strength data are expressed nearly by the power law of confining pressure. Such compilation of strength data and their plot have been given by a number of investigators (e.g., Handin, 1966).

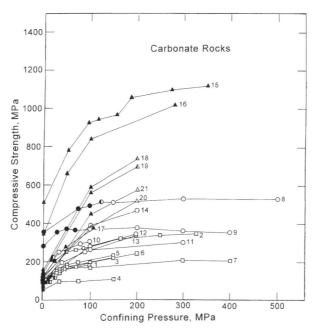

Figure 2.14.   Relation between compressive strength and confining pressure for carbonate rocks. Numbers given for each curve correspond to rocks listed in Table 2.2. Closed symbols: brittle behavior; semi-closed: transitional behavior; open: ductile behavior.

Table 2.2. A list of rocks.

No.	Rock	Tested by	Grouping
A. Silicate rocks			
1.	Aoishi Tuff	Mogi (1965)	
2.	Sakuishi Welded Tuff	Mogi (1965)	S_1 (Highly porous; Porosity $>10\%$)
3.	Shinkomatsu Andesite	Mogi (1965)	
4.	Mizuho Trachyte (1)	Mogi (1964)	
5.	Mizuho Trachyte (2)	Mogi (1965)	
6.	Tatsuyama Tuff (1)	Mogi (1964)	
7.	Tatsuyama Tuff (2)	Mogi (1965)	S_2 (Porous; Porosity: 1–10%)
8.	Shirochoba Andesite (1)	Mogi (1964)	
9.	Shirochoba Andesite (2)	Mogi (1965)	
10.	Bartlesville Sandstone	Handin & Hager (1957)	
11.	Barns Sandstone (1)	Handin & Hager (1957)	
12.	Westerly Granite	Mogi (1966)	
13.	Barre Granite	Robertson (1955)	
14.	Kitashirakawa Granite	Matsushima (1960)	
15.	Inada Granite	Mogi (1964)	
16.	Mannari Granite	Mogi (1965)	
17.	Nabeishi Peridotite	Mogi (1965)	
18.	Ukigane Diorite	Mogi (1964)	S_3^* (Nonporous; Porosity $<1\%$)
19.	Orikabe Diorite	Mogi (1965)	
20.	Cabramurra Serpentinite	Raleigh & Paterson (1965)	
21.	Tumut Pond Serpentinite	Raleigh & Paterson (1965)	
22.	Pala Gabbro	Serdengecti et al. (1961)	
23.	Frederic Diabase	Brace (1964)	
24.	Cheshire Quartzite	Brace (1964)	
B. Carbonate rocks			
1.	Wombeyan Marble	Paterson (1958)	
2.	Carrara Marble	Kármán (1911)	
3.	Yamaguchi Marble (Fine)	Mogi (1964)	
4.	Yamaguchi Marble (Coarse)	Mogi (1964)	CM (Marble)
5.	Mito Marble (1)	Mogi (1964)	
6.	Mito Marble (2)	Mogi (1965)	
7.	Danby Marble	Robertson (1955)	
8.	Solnhofen Limestone (1)	Heard (1960)	
9.	Solnhofen Limestone (2)	Robertson (1955)	
10.	Wells Station Limestone	Paterson (1958)	
11.	Becraft Limestone	Robertson (1955)	CL (Limestone)
12.	Devonian Limestone	Handin & Hager (1957)	
13.	Fusselman Limestone	Handin & Hager (1957)	
14.	Wolfcamp Limestone	Handin & Hager (1957)	
15.	Blair Dolomite (1)	Brace (1964)	
16.	Blair Dolomite (2)	Robertson (1955)	
17.	Webatuck Dolomite	Brace (1964)	
18.	Clear Fork Dolomite	Handin & Hager (1957)	CD (Dolomite)
19.	Fusselman Dolomite	Handin & Hager (1957)	
20.	Hasmark Dolomite (T)	Handin & Hager (1957)	
21.	Luning Dolomite	Handin & Hager (1957)	

Silicate rocks of No. 12–16 and No. 17–21 are denoted as S_3G and S_3P, respectively.

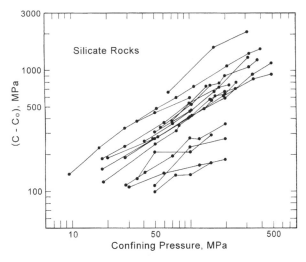

Figure 2.15.   Relation between (C–C$_0$) and confining pressure for silicate rocks, except for very porous rocks (S$_1$)). C: compressive strength, C$_0$: strength at atmospheric pressure.

However, most of these experiments have been carried out using the conventional test, which uses a short cylinder and is accompanied by stress concentration and the clamping effect at the end part of the specimen, as mentioned in Chapter 1. Therefore, compressive strength at atmospheric pressure and under low confining pressure should be measured by use of the revised specimen design. The experimental results for Westerly granite and Dunham dolomite which were obtained by the revised method are summarized in Table 2.3 and Figs. 2.16–2.17 (Mogi, 1966b). The larger circles in these figures represent the average strength of samples whose length/diameter ratio was greater than its critical value (2.5), so their reliability is high. The smaller circles were obtained from one or two experiments using specimens of length/diameter ratio 2.5, so their reliability is limited.

Fracture of Westerly granite was always very violent and explosive, and followed by a sudden stress drop under tested confining pressure (<400 MPa). It is noteworthy that the strength-pressure curve is not linear, but is concave downwards, especially at low pressure. This result seems to be inconsistent with some fracture criteria which predict a linear relation. In dolomite, strength increases nearly linearly with pressure.

Fracture angles in both rocks are shown in Fig. 2.16(b) and 2.17(b). Since the fracture angles are predicted from the Mohr envelope curve based on strength data, the observed fracture angles are also compared with the predicted angles by the Mohr envelope curves in these figures. In Dunham dolomite, the difference is small (0°–2.5°) and not systematic, so the Mohr theory would seem to apply here. In Westerly granite, the calculated curve is parallel to the observed curve and the calculated values are slightly lower than the observed values. The difference between them is (2°–4°) under lower confining pressure and decreases with pressure. The Mohr theory assumes that fracture occurs by shearing in a slip plane which depends on the value

Table 2.3.   Strength and fracture angle under different pressures.

Rock	Confining pressure (MPa)	Differential stress (MPa)	Observed fracture angle	Calculated fracture angle
Dunham dolomite	0.1*	209	20°	19° (?)
	8.2	289	20°	20.5°
	10.8	296	21°	21°
	18.0*	344	20°	22°
	21.6	367	22°	24.5°
	33.0*	412	25°	26° (?)
Westerly granite	0.1*	239	19°	–
	9.5	378	19.5°	15°
	17.0*	467	20.5°	17.5°
	27.5	555	22.5°	20°
	51.0*	720	24°	22°
	77.0	840	–	23°
	108.0*	977	26°	24.5°
	207.0	1308	–	27°
	321.0**	1597	32°	30.5°
	400.0**	1717	35°	34.5° (?)

* Average value from several tests.
** Average value from two tests.

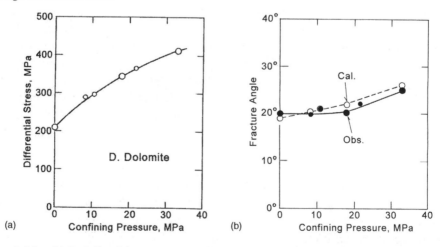

(a)   Confining Pressure, MPa

(b)   Confining Pressure, MPa

Figure 2.16.   (a) Relation between compressive strength and confining pressure for Dunham dolomite. (b) Relation between fracture angle and confining pressure for Dunham dolomite. Closed circle: observed angle; open circle: angle calculated from the Mohr envelope (Mogi, 1966b).

of the normal stress on the plane. The fracture in dolomite is a typical shear type. The fracture in granite specimens is also clearly of a shear type under high confining pressure, but under atmospheric or low pressure is often curved and quite irregular. The difference noted above between observation and calculation may be a manifestation

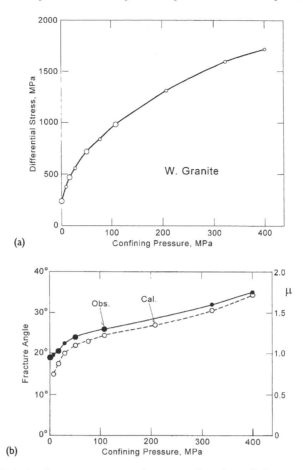

Figure 2.17. (a) Relation between compressive strength and confining pressure for Westerly granite (Mogi, 1966b). (b) Relation between fracture angle and confining pressure for Westerly granite. Closed circle: observed angle; open circle: angle calculated from the Mohr envelope.

of these irregularities. However, the difference (average 2.2°) between them may not be significant because the accuracy of angle measurement in granite is not very high.

From these precise measurements, it may be concluded that the Mohr theory applies rather well to shear fracture of a granite and dolomite in conventional triaxial compression ($\sigma_2 = \sigma_3$).

2.2.b The Coulomb-Mohr fracture criterion

As mentioned above, the compressive strength of brittle rocks is very dependent on confining pressure. Several fracture criteria have been proposed to correlate observed data. Among them, the Coulomb criterion which was introduced by Coulomb (1773), is the simplest and most widely applicable and used for prediction of the effect of confining pressure. The criterion states that shear fracture takes place when the shear stress

(τ) becomes equal to the sum of the cohesive shear strength (τ_0) and the product of the coefficient of internal friction (μ_i) and the normal stress (σ_n) across the fracture plane:

$$\tau = \tau_0 + \mu_i \sigma_n \tag{2.2}$$

If pore-fluid pressure exists, the role of normal pressure is supplanted by the effective normal pressure (Terzaghi, 1945). The Coulomb equation (2.2) is similar to the following equation for frictional sliding on external surfaces (Jaeger, 1959; Byerlee, 1967):

$$\tau = S_0 + \mu \sigma_n \tag{2.3}$$

where τ is the frictional resistance, σ_n is the normal stress, μ is the coefficient of external friction, and S_0 is a small constant. By comparison of Equations (2.2) and (2.3), it is suggested that μ_i in Equation (2.2) is similar to μ in Equation (2.3). The second term in the right hand side of Equation (2.2) was often interpreted as frictional resistance.

However, Handin (1969) emphasized that the physics of the Coulomb equation might be obscure. He noted that until the cohesion of rock is broken, no sliding surface exists in the intact body, and thus, the Coulomb equation seems to be contradictory, or the concept of internal friction is unsatisfactory. Thereafter, several investigators have also argued that the conventional obvious interpretation of the Coulomb fracture criterion cannot be correct because there is no slip on the fracture plane until after fracture (e.g., Scholz, 1990, p. 14; to some extent Paterson, 1978, p. 25).

On the other hand, the author (Mogi, 1974) proposed a very simple model for physical interpretation of the Coulomb fracture criterion. Recently, Savage, Byerlee and Lockner (1996) extended this model to include the Mohr fracture criterion. In the following, the Coulomb-Mohr fracture criterion is discussed from this point of view.

According to dilatancy and acoustic emission measurements which will be mentioned in later chapters, a number of micro-fractures occur during the deformational process prior to the macroscopic fracture. Fig. 2.18(A) shows schematically the locations of cracks along the eventual fault plane. To simplify the discussion, the area of eventual fault plane (F–F') just before the fracture is divided into the three parts, A_1, A_2, and A_3 [Fig. 2.18(B)]. A_1 is the area which is still intact, A_2 is the total area of closed crack surfaces, and A_3 is the total area of open crack surfaces. Since the open cracks have no effect in the shear resistance, the effect of A_3 can be ignored as the first step. Furthermore, it is assumed that the total shearing resistance (T) at the macroscopic shear fracture is the sum of the shearing resistance (T_1) in the intact part (A_1), which is independent of normal pressure, and the shearing resistance (T_2) in the closed cracks (A_2), which is simply assumed to be the frictional resistance. Therefore the shearing stress (τ) at fracture is:

$$\tau = T/A$$
$$= T_1/A + T_2/A \tag{2.4}$$
$$= \tau_1(A_1/A) + \mu\sigma_n(A_2/A) \tag{2.5}$$

where τ_1 is the inherent shearing strength of the intact part which is independent of pressure and μ is the coefficient of external friction. From the comparison of

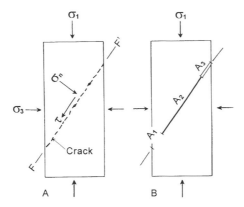

Figure 2.18.   Simplified models of the cracked state along the eventual fault plane just before faulting.

Equations (2.2) and (2.5):

$$\tau_0 = \tau_1(\mathbf{A}_1/\mathbf{A}) \tag{2.6}$$

$$\mu_i = \mu(\mathbf{A}_2/\mathbf{A}) \tag{2.7}$$

These relations suggest a physical model for the cohesive strength (τ_0) and the coefficient of internal friction (μ_i).

According to the observation of microfracturing and the estimation from the strength versus pressure curve, the total area of closed crack surface ($\mathbf{A}_2$) just before the macroscopic fracture probably accounts for a large portion of the fault plane area ($\mathbf{A}$). Consequently, the coefficient of the internal friction (μ_i) is nearly equal to, although not the same as the coefficient of the external friction (μ). The cohesive strength (τ_0) is much smaller than the inherent shearing strength (τ_1) of the perfectly intact part. For example, $\mathbf{A}_2/\mathbf{A}$ and τ_0/τ_1 for Westerly granite were estimated from the strength versus pressure curve at 7/8 and 1/8, respectively, as will be mentioned in the later part. This is a model proposed by the author for explanation of the physical meaning of the Coulomb equation (Mogi, 1974).

Some investigators (e.g., Brace, 1964; Hoek, 1965) argued that there is good agreement between observed fracture strength of certain brittle rocks and prediction of the Coulomb criterion, taking the coefficient of internal friction 0.5–1.5; some of these values were, however, higher than the coefficients of sliding friction. Also, in experimental results under a wider pressure range, the strength versus pressure curves of some rocks are markedly concave towards the pressure axis at low confining pressure and near the brittle-ductile transition pressure. Therefore, the Coulomb criterion does not always hold for a wide pressure range, and the strength versus pressure curves are rather expressed by a power function (Matsushima, 1960; Murrell, 1965; Mogi, 1966; Byerlee, 1969).

In the following, it is shown that the pressure effect on the strength of brittle rocks is fundamentally expressed by the Coulomb theory, except for a low pressure region

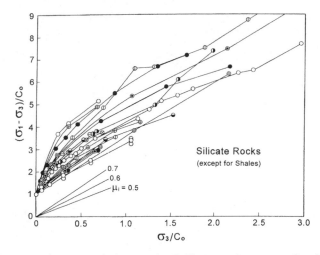

Figure 2.19.   Compressive strength ($\sigma_1 - \sigma_3$) of silicate rocks, except for shales, against the confining pressure ($p = \sigma_2 = \sigma_3$), normalized by the uniaxial compressive strength C_0. μ_i: coefficient of internal friction.

in some rocks. Fig. 2.19 shows the strength versus pressure curves for silicate rocks (except for shales), based on published data until 1974 (Robertson, 1955; Handin and Hager, 1957; Matsushima, 1960; Brace, 1964; Mogi, 1964, 1965; Murrell, 1965; Raleigh and Paterson, 1965; Franklin and Hoek, 1970; Hoshino et al., 1972). These curves are limited to a very brittle region and the high-pressure region near the brittle-ductile transition pressure is not included. In this figure, the compressive strength and the confining pressure are normalized by dividing by the uniaxial compressive strength C_0. In this coordinate system, the Coulomb equation is expressed by:

$$(\sigma_1 - \sigma_3)/\mathbf{C}_0 = 1 + \alpha\sigma_3/\mathbf{C}_0 \qquad (2.8)$$

and the coefficient (α) is related with the coefficient of internal friction (μ_i) by:

$$\alpha = \left[(\mu_i^2 + 1)^{1/2} + \mu_i\right] / \left[(\mu_i^2 + 1)^{1/2} - \mu_i\right] - 1 \qquad (2.9)$$

In Fig. 2.19, some of observed strength versus pressure curves are nearly linear and parallel to each other. Other curves, however, are concave toward the pressure axis in a low-pressure range and nearly linear in a higher pressure range. It is noticeable that the linear parts in both cases are roughly parallel to each other. The average slope (α) of the linear part of the strength versus pressure curves of these silicate rocks (except for shales) corresponds to a coefficient of internal friction of about 0.6. According to recent friction measurements, the coefficients of sliding friction of various silicate rocks are nearly equal to the above-mentioned coefficient of internal friction.

The coefficient of sliding friction of shales is significantly smaller than that of other silicate rocks (Maurer, 1965). If the Coulomb fracture criterion is applicable, it is expected that the slope (α) of the strength versus pressure curves of shales is smaller

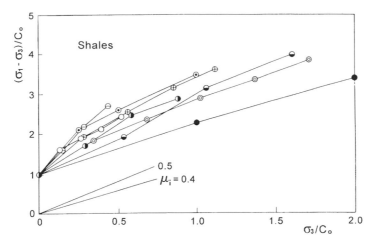

Figure 2.20. Compressive strength $(\sigma_1 - \sigma_3)$ of shales against the confining pressure $(p = \sigma_2 = \sigma_3)$, normalized by the uniaxial compressive strength C_0. μ_i: coefficient of internal friction.

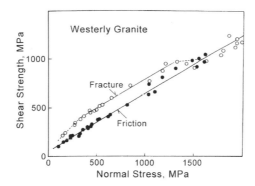

Figure 2.21. Shear stress versus normal stress at fracture (open circles) and friction on mated surfaces (solid circles) for Westerly granite. Data from Byerlee (1967).

than that of other silicate rocks. Fig. 2.20 shows that this is true. The coefficient of internal friction (about 0.4–0.5) obtained from the average slope of the linear part seems to be roughly equal to the coefficient of sliding friction (μ) obtained from Maurer's experimental result.

Figure 2.21 also supports the close relation between the fracture and friction curves. This figure was drawn by the author (Mogi, 1974) using the data from Byerlee (1967). In this figure, the following features can be pointed out: In Westerly granite, the shear stress versus normal pressure curve for fracture is nearly linear and parallel to the curve for sliding friction, except for the low-pressure region ($\sim$400 MPa) and high-pressure region ($\sim$1300 MPa). At the pressure of 1500 MPa and higher, the fracture curve agrees with the friction curve, and so deviation from the linear relation near

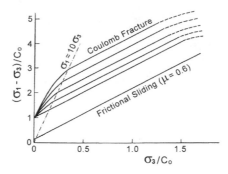

Figure 2.22. Typical strength versus pressure curves of rocks and the boundary line $(\sigma_1 \approx 10\sigma_3)$ between the regions where the Coulomb criterion holds and where it does not hold.

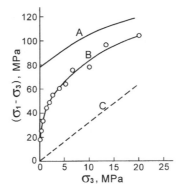

Figure 2.23.   Compressive strength $(\sigma_1 - \sigma_3)$ as a function of confining pressure $(p = \sigma_2 = \sigma_3)$ for a marble. Curve A: original marble; curve B: marble with grain boundaries disintegrated by heating (after Rosengren and Jaeger, 1968); straight line C: frictional sliding $(\mu = 0.6)$.

the normal pressure of 1500 MPa is attributed to the brittle-ductile transition process, which will be discussed later.

Thus, the Coulomb fracture criterion appears to apply to brittle fracture of many rocks, except for a low-pressure region and near the brittle-ductile transition pressure. Fig. 2.22 shows schematically the strength-pressure curves of various rocks. The curves of some rocks are concave toward the pressure axis in a low-pressure region. This high pressure-sensitivity of strength cannot be explained as a direct result of the pressure-sensitivity of the coefficient of sliding friction (Byerlee, 1967; Handin, 1969).

An interesting experimental result by Rosengren and Jaeger (1968) suggests a possible interpretation of high pressure-sensitivity of strength of rocks at low confining pressure. Their result is shown in Fig. 2.23. Curves A and B are the compressive fracture strength versus confining pressure curves for a virgin marble and for the same marble with grain boundaries disintegrated by heating, respectively. A curve for common sliding friction of marble is schematically shown by straight line C. It is noticeable that curve B for the highly cracked marble is greatly different from curves A and C. This difference may be explained by the following two processes. The first is that the shear resistance for sliding on highly rough and interlocked surfaces, such as grain boundaries or joints, is much higher than that on smooth surfaces under low confining pressure. The second is that a number of open cracks at atmospheric pressure close with increase of confining pressure and so A_2 in Equation (2.5) or (2.7) increases, that is, the shear resistance of closed cracks increases markedly under confining pressure. Thus, it is reasonable that the high pressure-sensitivity of strength of the disintegrated marble is attributed to such effects of the cracked state. The strength versus pressure curves in the low pressure-region of silicate rocks shown in Figs. 2.19 and 2.20 can also be explained similarly by the state of cracks in rocks. For a high crack porosity, the pressure sensitivity of fracture strength is very high under low confining pressure and decreases with increasing confining pressure, and so the strength versus pressure curve tends to be concave toward the pressure axis at low confining pressure. In very compact rocks without appreciable pre-existing open cracks, the strength-pressure curve is nearly straight. In this case, the Coulomb fracture criterion is applicable. As mentioned above, in cases where there are appreciable open cracks, the strength versus pressure curve is concave toward the pressure axis and the Mohr criterion can be applied.

Recently, Savage, Byerlee and Lockner (1996) discussed this subject based on their careful measurements on fracture and friction of Westerly granite under confining pressure, from a similar standpoint as the author's. They obtained A_1/A, A_2/A and A_3/A as functions of the normal stress across that surface at failure and concluded that "the curvature in the strength-versus-normal-stress curve is explained in the Mogi model as being caused by changes in the relative proportions of cracked and intact areas on the incipient rupture surface, proportions which depend upon the prefailure loading trajectory." They state that what is called internal friction is simply a manifestation of ordinary friction. Moreover, the curvature of the Mohr failure envelope is a consequence of changes in the incipient rupture surface occupied by closed cracks, and that the portion depends in turn upon damage accumulated along the stress trajectory used to reach failure. These results of Mogi (1974) and Savage et al. (1996) showed that the Coulomb-Mohr criterion is not strictly empirical, but that it can be explained by a reasonable physical model.

2.2.c *Brittle-ductile transition*

With increase of confining pressure, ductility, which is defined as the ability to undergo large permanent deformation without fracture (Handin, 1966), increases markedly and a transition from the brittle to the ductile state takes place at some confining pressure. Fig. 2.24 shows typical stress-strain curves in the brittle, the transitional and the ductile state. Figs. 2.25 and 2.26 show the brittle-ductile behavior in the conventional triaxial compression test as functions of the confining pressure and compressive strength of silicate rocks and carbonate rocks, respectively (Mogi,

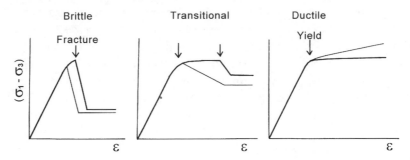

Figure 2.24.   Typical stress-strain curves in the brittle, the transitional and the ductile state.

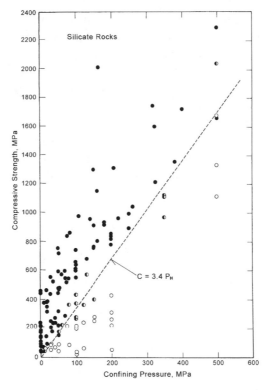

Figure 2.25.   Failure behavior of silicate rocks at various strength and pressure. Dotted line: boundary between brittle region and ductile region; closed circle: brittle; semi-closed circle: transitional; open circle: ductile.

1966). In silicate rocks, the brittle state (closed circles) region and the ductile state (open circles) region is clearly divided by a straight line passing through the origin (Fig. 2.25). This boundary line is expressed by

$$(\sigma_1 - \sigma_3) = 3.4\sigma_3 \tag{2.10}$$

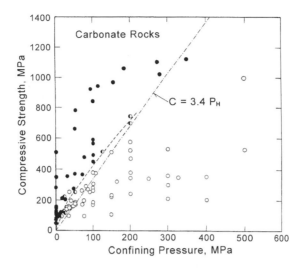

Figure 2.26. Failure behavior of carbonate rocks at various strength and pressure. Dotted line: boundary between brittle region and ductile region; closed circle: brittle; semi-closed circle: transitional; open circle: ductile.

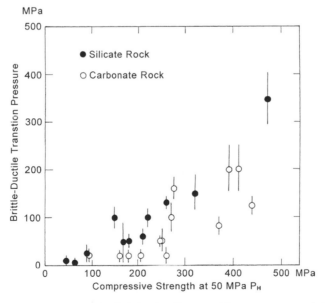

Figure 2.27. Relation between the brittle-ductile transition pressure and the compressive strength at 50 MPa confining pressure.

On the other hand, the boundary for carbonate rocks is somewhat different from that of silicate rock and clearly shifts to left side (Fig. 2.26). Fig. 2.27 shows the relation between the brittle-ductile transition pressure and the compressive strength at 50 MPa confining pressure in silicate rocks (closed circles) and carbonate rocks (open circles).

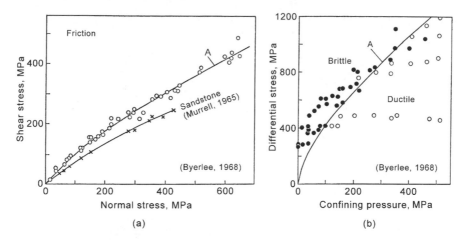

Figure 2.28.   (a) Shear stress versus normal stress for friction. Circles: six rocks including Solnhofen limestone and Westerly granite (Byerlee, 1968); crosses: Darley Dale sandstone (Murrell, 1965). (b) Brittle-ductile behavior of five rocks as a function of confining pressure and differential stress at fracture or 5% strain if the specimen was ductile. Curve A is the boundary between brittle and ductile regions determined from friction data (Byerlee, 1968).

The brittle-ductile transition pressures of silicate rocks are appreciably higher than those of carbonate rocks. This difference between silicate rocks and carbonate rocks suggests that there are different mechanisms of the brittle-ductile transition in different rock types.

Orowan (1960) suggested that the stress drop characteristic of fracture does not occur at high pressure because frictional resistance on the fault surface becomes higher than the shearing strength of rock. This idea has been used as a possible explanation for the brittle-ductile transition by Maurer (1965), Mogi (1966), Byerlee (1968) and others. As mentioned above, the author (Mogi, 1966) pointed out that the brittle-ductile transition boundary in the strength versus pressure graphs is expressed by a nearly linear curve for various silicate rocks and the transition mechanism may be explained by Orowan's frictional hypothesis. The transition boundary in carbonate rocks is somewhat different from that in silicate rocks, this being attributed to a different transition mechanism. Byerlee (1968) discussed this problem based on his measurement of friction of rocks, and he argued that the brittle-ductile transition boundary is independent of rock type, and that Orowan's frictional hypothesis is applicable for both silicate and carbonate rocks. Fig. 2.28 shows the result by Byerlee (1968).

The author (Mogi, 1972) suggested that the frictional sliding hypothesis is applicable for the brittle-ductile transition process of rocks (noted as B-type) in which the permanent deformation in the post-yield region occurs by cataclastic flow or frictional sliding, but that it may not be applied to those of A-type, in which large permanent strain before fracture occurs by homogeneous plastic deformation. The typical stress-strain curves of A- and B-types are schematically shown in Figs. 2.29 and 2.30,

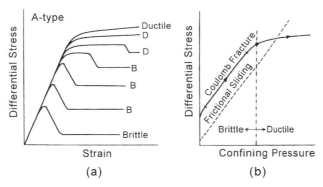

Figure 2.29.   (a) Typical stress-strain curves of A-type rocks for different confining pressures. (b) Strength versus pressure curve and the failure behavior in A-type rocks.

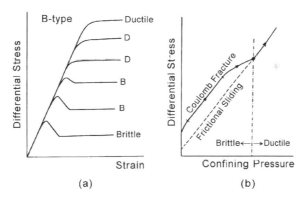

Figure 2.30.   (a) Typical stress-strain curves of B-type rocks for different confining pressures. (b) Strength versus pressure curve and the failure behavior in B-type rocks.

respectively. Some carbonate rocks, particularly at high temperature, are A-type and silicate rocks are B-type.

Thus, the pressure dependence of strength of rocks near the transition pressure is different between A- and B-types. The strength versus pressure curves and their possible interpretations are summarized as follows.

1. *A-type*

The typical stress-strain curves of A-type rocks for different confining pressures and the strength versus pressure curve are shown in Fig. 2.29. In this case, faulting occurs after some permanent deformation, which increases with increasing pressure. The large permanent strain in the post-yield region before fracture occurs by homogeneous plastic deformation. The brittle-ductile transition may take place when the fracture strength is equal to the yield strength. In Fig. 2.29(b), the strength in a brittle region increases nearly linearly with increasing pressure and then the curve approaches the brittle-ductile transition boundary while at the same time decreasing its slope. The decrease in the slope of curve near the transition pressure may be attributed to the gradual increase of local yielding due to heterogeneity (Mogi, 1966). At higher pressure, the yield strength approaches a constant value.

2. *B-type*

The typical stress-strain curves in this case are shown in Fig. 2.30 (a). The stress drop occurs just after the yield point and the permanent deformation in the post-yield region occurs by cataclastic flow or frictional sliding. The brittle-ductile transition pressure is the pressure at which the strength of rock at faulting is equal to the strength due to frictional resistance after faulting (Orowan, 1960). Figure 2.30 (b) shows the typical strength versus pressure curve. This curve is nearly parallel to the curve for frictional sliding on the fault surface, except for a low-pressure region. The slope of curve decreases gradually near the brittle-ductile transition pressure. This gradual decrease in the slope of curve is attributed to the increase of local fractures before faulting due to heterogeneity. That is, the area of the intact part (A_1) in Equation (2.6) decreases gradually near the transition pressure, at which A_1 becomes zero. At pressures higher than the transition pressure, the strength versus pressure curve agrees with the curve for frictional sliding, in which the strength increases linearly with increasing pressure. The strength versus pressure relation of Westerly granite shown in Fig. 2.21 is typical for the B-type.

Most rocks, however, behave in a manner intermediate between A- and B-types. Near the transition pressure, inelastic deformation just before and after yielding probably occurs both by fracturing and plastic deformation.

REFERENCES

Adams, F. D. and J. T. Nicolson. (1901). An experimental investigation into the flow of marble, Royal Soc. London Philos. Trans., Ser. A, **195**, 597–637.

Birch, F. (1960). The velocity of compressional waves in rocks to 10 kilobars, Part 1. J. Geophys. Res., **65**, 1083–1102.

Böker, R. (1915). Die Mechanik der bleibenden Formänderung in Kristallinisch aufgebauten Körpern. Ver. Dtsch. Ing. Mitt. Forsch., **175**, 1–51.

Brace, W. F. (1964). Brittle fracture of rocks. In: State of Stress in the Earth's Crust. Judd, W. R. (ed.). New York, Elsevier, 111–174.

Byerlee, J. D. (1967). Frictional characteristics of granite under high confining pressure. J. Geophys. Res., **72**, 3639–3648.

Byerlee, J. D. (1968). Brittle-ductile transition in rocks. J. Geophys. Res., **73**, 4741–4750.

Coulomb, C. A. (1773). Sur une application des régles maximis et minimis á quelques problémes de statique, relatifs á l'architecture. Acad. Sci. Paris Mem. Math. Phys., **7**, 343–382.

Franklin, J. A. and E. Hoek. (1970). Developments in triaxial testing technique. Rock Mech., **2**, 223–228.

Griggs, D. T. (1936). Deformation of rocks under confining pressure. J. Geol., **44**, 541–577.

Handin, J. (1966). Strength and ductility. In: Handbook of Physical Constants, Revised ed., Clark, S. P. (ed.), Geol. Soc. Am. Memoir, **97**, 223–289.

Handin J. (1969). On the Coulomb-Mohr failure criterion. J. Geophys. Res., **74**, 5343–5350.

Handin, J. and R. V. Hager. (1957). Experimental deformation of sedimentary rocks under confining pressure: tests at room temperature on dry samples. Bull. Am. Assoc. Petrol. Geol., **41**, 1–50.

Heard, H. C. (1960). Transition from brittle fracture to ductile flow in Solenhofen limestone as a function of temperature, confining pressure, and interstitial fluid pressure. In: Rock Deformation. Griggs, D., Handin, J. (eds.) Geol. Soc. Am., Memoir, **79**, 193–226.

Heard, H. C. (1963). Effect of large changes in strain rate in the experimental deformation of Yule marble. J. Geol., **71**, 162–195.

Hoshino, K., H. Koide, K. Inami, S. Iwamura and S. Mitsui. (1972). Mechanical properties of Japanese tertiary rocks under high confining pressure. Geol. Surv. Japan Rep., **244**, 1–200.

Jaeger, J. C. (1959). The frictional properties of joints in rock. Geofis. Pura Appl., **43**, 148–158.

Kármán, T. von (1911). Festigkeitsversuche unter allseitigem Druck. Z. Verein. Dtsch. Ing. **55**, 1749–1757.

Matsushima, S. (1960). On the deformation and fracture of granite under high confining pressure. Bull. Disaster Prevention Res. Inst. Kyoto Univ., **36**, 11–20.

Maurer, W. C. (1965). Shear failure of rock under compression. Soc. Pet. Eng. J., **5**, 167–176.

Mogi, K. (1964). Deformation and fracture of rocks under confining pressure (1) Compression tests on dry rock sample. Bull. Earthquake Res. Inst., Tokyo Univ., **42**, 491–514.

Mogi, K. (1965). Deformation and fracture of rocks under confining pressure (2), Elasticity and plasticity of some rocks. Bull. Earthquake Res. Inst., Tokyo Univ., **43**, 349–379.

Mogi, K. (1966a). Pressure dependence of rock strength and transition from brittle fracture to ductile flow. Bull. Earthquake Res. Inst., Tokyo Univ., **44**, 215–232.

Mogi, K. (1966b). Some precise measurements of fracture strength of rocks under uniform compressive stress. Felsmechanik und Ingenieurgeologie, **4**, 41–55.

Mogi, K. (1972). Fracture and flow of rocks. In: A. R. Ritsema (ed.), The Upper Mantle. Tectonophysics, **13** (1–4), 541–568.

Mogi, K. (1974). On the pressure dependence of strength of rocks and the Coulomb fracture criterion. Tectonophysics, **21**, 273–285.

Murrell, S. A. F. (1965). The effect of triaxial stress systems on the strength of rocks at atmospheric temperatures. Geophys. J. R. Astron. Soc., **10**, 231–281.

Orowan, E. (1960). Mechanism of seismic faulting. Geol. Soc. Am. Mem., **79**, 323–345.

Paterson, M. S. (1958). Experimental deformation and faulting in Wombeyan marble. Bull. Geol. Soc. Am., **69**, 465–476.

Paterson, M. S. (1978). Experimental Rock Deformation – The Brittle Field. Springer-Verlag, Berlin, pp. 254.

Raleigh, C. B. and M. S. Paterson. (1965). Experimental deformation of serpentinite and its tectonic implications. J. Geophys. Res., **70**, 3965–3985.

Robertson, E. C. (1955). Experimental study of the strength of rocks. Bull. Geol. Soc. Am., **66**, 1275–1314.

Rosengren, K. J. and J. C. Jaeger. (1968). The mechanical properties of an interlocked low-porosity aggregate. Geotechnique, **18**, 317–326.

Savage, J. C., J. D. Byerlee and D. A. Lockner. (1996). Is internal friction friction? Geophys. Res. Let., **23**, No. 5, 487–490.

Serdengecti, S. and G. D. Boozer. (1961). The effects of strain rate and temperature on the behavior of rocks subjected to triaxial compression. In: Proc. 4th Symp. Rock Mechanics. Hartman, H. L. (ed.). Penn. State Univ., Bull. Min. Ind. Exp. Sta., No. 76, 83–97.

Scholz, C. H. (1990). Mechanics of Earthquakes and Faulting. Cambridge Univ. Press, Cambridge, pp. 439.

Terzaghi, K. (1945). Stress conditions for the failure of saturated concrete and rock. Proc. Am. Soc. Test. Mater., **45**, 777–801.

Yamamoto, K., Y. Kuwahara, N. Kato and T. Hirasawa. (1990). Deformation rate analysis: A new method for *in situ* estimation from inelastic deformation of rock samples under uni-axial compressions. Tohoku Geophys. Journ., **33**, No. 2, 127–147.

CHAPTER 3

Deformation and fracture of rocks under the triaxial compression: The effect of the intermediate principal stress

3.1 HISTORY OF COMPRESSION EXPERIMENTS

The mechanical properties, such as failure strength, fracture angle, and ductility, are function of state of stress, temperature and strain rate (e.g. Paterson and Wong, 2005). In this book, only the effects of stress state are discussed. As mentioned in the preceding chapter, the stress state can be specified by the three principal stresses, σ_1 (the maximum principal stress), σ_2 (the intermediate principal stress) and σ_3 (the minimum principal stress) (compressive stress condition is taken positive).

Since the pioneer work of Adams and Nicolson (1901), there have been many experimental studies of deformation and failure of rocks under combined stresses. The main experiments of rock deformation under combined stresses are shown as time series in Table 3.1. The principal points of the methods of these experiments and important remarks related to experimental accuracy and stress level, are concisely

Table 3.1. History of triaxial experiments on rocks.

Date	Researcher	Stress system	Method	Remarks
1901	Adams & Nicolson	$\sigma_1 \geq \sigma_2 = \sigma_3$	σ_1: solid piston $\sigma_2 = \sigma_3$: solid pipe	{ low accuracy by friction
1911	Kármán	$\sigma_1 \geq \sigma_2 = \sigma_3$	σ_1: solid piston $\sigma_2 = \sigma_3$: fluid pressure	{ high accuracy high stress
1915	Böker	$\sigma_1 = \sigma_2 \geq \sigma_3$	$\sigma_1 = \sigma_2$: fluid pressure σ_3: solid piston	{ high accuracy high stress
1935	Griggs	$\sigma_1 \geq \sigma_2 = \sigma_3$	same as Kármán σ_3: constant pressure	{ high accuracy high stress
1967	Handin et al.	$\sigma_1 \geq \sigma_2 \geq \sigma_3$	hollow cylinder specimen solid piston (compression + torsion) + fluid pressure	{ low accuracy
1968	Hojem & Cook	$\sigma_1 \geq \sigma_2 \geq \sigma_3$	σ_1: solid piston σ_2, σ_3: thin flat jack (nearly fluid pressure)	{ high accuracy low stress few data
1969*	Mogi	$\sigma_1 \geq \sigma_2 \geq \sigma_3$	σ_1, σ_2: solid piston σ_3: fluid pressure	{ high accuracy high stress
1977	Mogi	$\sigma_1 \geq \sigma_2 \geq \sigma_3$	σ_1: solid piston σ_2, σ_3: fluid pressure	{ high accuracy high stress

* see Appendix

listed. For quantitative discussions, the experimental accuracy is essential. The possibility of experiments under high stress conditions is also very important, because the application of high stress is indispensable for investigation of the mechanical behavior of rocks at great depth.

3.1.a   *Axial loading test under lateral pressure*

Adams and Nicolson (1901) applied an axial compressive stress to a cylindrical rock specimen which was covered by a metal pipe. With the increase of lateral dilatation under axial compression, the lateral compressive stress is applied to the rock specimen. They found the following effects of the lateral pressure: (1) marked increase of failure strength and (2) appreciable increase of ductility. These two most important effects of the lateral pressure were pointed out clearly by this classical experiment. However, these results were qualitative because the quantitative estimation of the lateral pressure was difficult and the effect of friction between the rock sample and the metal pipe was complex.

The quantitative triaxial experiments date from the work of von Kármán (1911). In these tests, σ_1 is applied to a cylindrical rock specimen by an axial solid piston, and the lateral stress is applied by the fluid-confining pressure. The Kármán-type triaxial testing machine has been used widely without any essential change by many investigators. As mentioned in the preceding chapter, the end effect of axial compression markedly decreases with increase of the confining pressure. Therefore, the stress distribution in the rock sample is quite uniform and the precise measurements of stress and strain can be carried out. By using a high pressure vessel, high stresses can be applied without any technical difficulties. In this usual triaxial compression test, the stresses are axisymmetric ($\sigma_1 > \sigma_2 = \sigma_3$).

Böker (1915) did experiments in which the confining pressure was greater than the axial stress, so that $\sigma_1 = \sigma_2 > \sigma_3$. This is the triaxial extension test. Thereafter, a number of investigators carried out experiments of this type and often discussed whether or not the relative value of σ_2 influences the fracture strength comparing of the triaxial compression and triaxial extension test results. This topic is discussed in the next section.

3.1.b   *True triaxial compression test*

As mentioned above, in these conventional triaxial tests of the Kármán-type, two principal stresses are equal ($\sigma_2 = \sigma_3$ or $\sigma_1 = \sigma_2$). Since around 1960, a number of experiments have been made for studies of the role of all three principal stresses in failure strength and ductility of rocks. Figure 3.1 schematically shows the various methods (A), (B), (C) and (D) devised for the true triaxial compression test in which all three principal stresses are different, $\sigma_1 \geq \sigma_2 \geq \sigma_3 > 0$. The thick arrows indicate compression or torsion through solid piston or end pieces and the thin arrows indicate compression by fluid pressure.

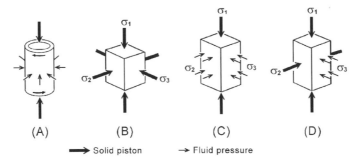

(A)   (B)   (C)   (D)

⟹ Solid piston   → Fluid pressure

Figure 3.1. Various methods devised to test rock samples under conditions in which all three principal stresses are different.

- *A-type experiment: Compression and torsion of a hollow cylinder under fluid confining pressure*

Previously a thick hollow cylinder was used in this type of experimental arrangement (for example, Robertson, 1955; Hoskins, 1969). In this case, stress distribution is inhomogeneous and stress gradient exists in rock specimens, so the significance of the experimental results is not clear.

In order to achieve nearly homogeneous stress distribution, Handin, Heard and Magouirk (1967) used a very thin hollow cylinder. By axial compression and torsion of a thin hollow cylinder under fluid-confining pressure, different three principal stresses can be applied. This method is similar to the experimental arrangement in ductile metals. Their paper has been referred widely as an important research on the σ_2-effect in rocks (for example, Paterson, 1978), without any suitable criticism. However, according to the author's opinion, their experimental method has a vital defect. In Fig. 3.2 (right), experimental data listed in their paper are plotted in the σ_1–σ_2 graph. Different symbols indicate different σ_3 values. In this figure, data greatly scatter, and so any significant regularity cannot be recognized. In the left figure, our experimental results reported in 1969 (Mogi, 1970a,b) (see later section) are shown. These two graphs are quite different.

Such marked scattering or very low reproducibility of experimental data may be attributed to use of a very thin hollow cylindrical specimen. The wall thickness in this experiment is only 0.7 mm. In fabrication of such thin hollow cylinder of brittle rock (Solnhofen limestone), generation of micro-cracks is inevitable. This situation of brittle rocks is quite different from that of ductile metals. Such cracks of the thin hollow cylinder give serious effects in strength measurement. As far as ductility is concerned, the statement in that paper that ductility is observed to increase with mean pressure is also unsuitable (see later section (3.3 f)).

In conclusion, the method of Handin, Heard and Magouirk (1967) is unsuitable for the true triaxial test.

- *B-type experiment: Triaxial compression by three solid pistons*

There are many experiments in which three opposite pairs of faces of cubic specimens are compressed by solid pistons (for example, Niwa, Kobayashi and Koyanagi,

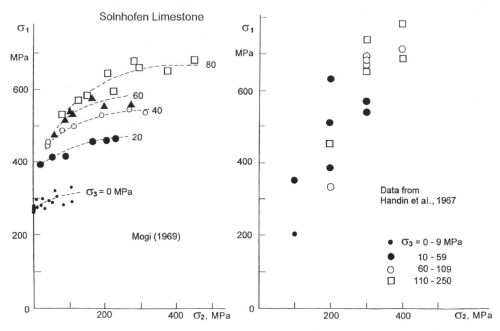

Figure 3.2.   Right figure: Experimental results on the relation between σ_1 and σ_2 for Solnhofen limestone reported by Handin et al. (1967) (Data of Table 6 were plotted by Mogi). Left figure: Similar figure by Mogi (1969).

1967). In these cases, special precautions are taken to minimize frictional constraint at the loading platens. However, the stress distribution in rock specimens is not homogeneous using this method, as discussed in the Chapter 1.1. Therefore, the precise true triaxial test is difficult using this experimental arrangement.

- *C-type experiment: Axial compression (σ_1) by a solid piston and lateral compression (σ_2 and σ_3) by two different fluidal pressures*

Figure 3.3 shows the triaxial cell developed by Hojem and Cook (1968). σ_1 is applied by an axial solid piston, and σ_2 and σ_3 are applied independently by two pairs of copper flat-jacks. Since the flat jacks are soft, the lateral pressure seems to be nearly similar to the fluidal pressure. Figure 3.4 shows the fracture stress (σ_1) of Karroo dolerite as a function of σ_2 obtained by Hojem and Cook (1968). This result shows that the σ_2-effect is appreciable, although it is much smaller than the σ_3-effect. Although there is no direct evidence to show the homogeneous stress distribution, this experiment seems to be a reliable true triaxial compression test. In this method, however, the application of high lateral stresses is difficult because of the limited strength of thin soft copper membrane in the flat-jack. In Fig. 3.4, σ_2 and σ_3 are lower than 50 MPa. Moreover, the number of data in their paper are very small and they did not continue this experiment further.

Mogi (1977) carried out an experiment in which σ_1 is applied by an axial solid piston, σ_3 is applied by the fluid-confining pressure, and σ_2 is applied by pressure

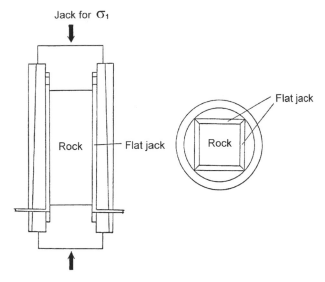

Figure 3.3.   Triaxial cell by Hojem and Cook (1968). σ_1 is applied by solid piston, and σ_2 and σ_3 are applied by two flat jacks.

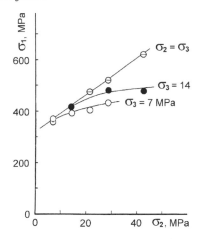

Figure 3.4.   Experimental result by Hojem and Cook (1968). The fracture stress (σ_1) of Karroo dolerite as a function of σ_2.

using very soft rubber. This experiment which has features of the D-type method was carried out to check this method. It is explained in detail in the following section (Chapter 3: Fig. 3.18).

- *D-type experiment: Triaxial compression by two pairs of solid pistons (σ_1 and σ_2) and fluid-confining pressure (σ_3)*

This true triaxial compression method was developed by Mogi (1969, 1970a). This method was achieved by an important modification of the Kármán-type triaxial testing apparatus. σ_1 is applied by the axial solid piston and σ_2 is applied by a lateral piston, and σ_3 is applied by the fluid-confining pressure. The minimum principal stress (σ_3)

higher than 800 MPa can be applied by use of the high pressure vessel with thick wall. As will be mentioned in the later section, the homogeneous true triaxial compression test in the wide range of stresses is possible using this experimental arrangement. Since 1969, a number of experiments have been carried out on 7 rocks including weak and hard rocks. By a series of experiments, the effects of the intermediate principal stress (σ_2) to failure strength, fracture angle, ductility and dilatancy have been investigated, and the empirical failure criteria have been proposed. These results including unpublished materials are fully presented in this chapter.

In 1979 the triaxial compression testing machine by the above-mentioned principle was on sale as "Mogi-type true triaxial testing machine" by Marui Company, Osaka, Japan. In China, Xu (1980) constructed "the Mogi-type triaxial machine" and his group has carried out various experiments on the σ_2-effect in rock fracture strength by use of this machine. Takahashi and Koide (1989) carried out "the true triaxial compression experiments" on soft rocks such as sandstones. However, their triaxial apparatus is significantly different from the "Mogi-type machine", as mentioned in Section 3.3. Recently, Wawersik et al. (1997) and Haimson and Chang (2000) con-structed the true triaxial cells which emulate Mogi's original design. In recent years, the importance of the σ_2-effect problem is seen in a new light, and these experimental data are discussed by various investigators (e.g., Colmenares and Zoback, 2002; Fjær and Ruistuen, 2002; Al-Ajmi and Zimmerman, 2005).

Recently, Chang and Haimson (2000 a) wrote as follows:

"A number of previous studies have shown that the intermediate principal stress possesses a strengthening effect on rock strength. However, the true triaxial test-ing devices used relatively low loading capacities and *were not suitable for testing the compressive strength of hard rocks under high true triaxial stress conditions.* A direct result of the difficulty to design and fabricate intricate true triaxial apparatus (Wawersik et al., 1997), and to conduct a long and tedious series of rather complicated experiments, has been a scarcity of such testing. *The few published results were inter-preted more as interesting curiosities, rather than serious challenges to the accepted Mohr-type criteria.*" (italic by the present author).

If the above-mentioned review of the history of triaxial compression experiments and the following explanation in this chapter are considered, this statement is clearly unsuitable. Very recently, Al-Ajimi and Zimmerman (2005) wrote that we propose that the linear version of the Mogi criterion be known as the "Mogi-Coulomb" failure criterion.

3.2   COMPARISON BETWEEN COMPRESSION AND EXTENSION UNDER CONFINING PRESSURE

3.2.a   *Introduction*

As mentioned in the preceding section, since the pioneer works of von Kármán (1911) and Böker (1915) on the triaxial compression and extension, there have been many experimental studies of the fracture of rocks under combined stresses. According to the triaxial compression test of the Kármán-type ($\sigma_1 > \sigma_2 = \sigma_3$), fracture strength of brittle rocks markedly increases with increasing σ_3. This σ_3-effect is well explained by the Coulomb, Mohr, Griffith, and modified Griffith theories, which do not consider the effect of the intermediate principal stress σ_2. Böker (1915) compared the result of extension test ($\sigma_1 = \sigma_2 > \sigma_3$) on Carrara marble with that of compression test on the same rock by von Kármán. Böker found a slight difference between them, which has been attributed by many investigators to the influence of the intermediate principal stress. This conclusion may be not certain, because there is a possibility that the difference could have been due to anisotropy of the rock.

Thereafter, some experimenters (for example, Handin and Hager, 1957; Murrell, 1965; Handin, Heard and Magouirk, 1967) also found a discrepancy between the Mohr envelopes for compression and extension under confining pressure; others (for example, Brace, 1964) could find no significant difference. The present author found that this disagreement between two groups may be attributed to the low accuracy of failure stress measurements, particularly in extension tests. The influence of σ_2 may be relatively small; very careful and precise measurements are therefore essential.

From this standpoint, the following careful experiment was carried out to study the effect of the intermediate principal stress on rock failure (Mogi, 1967).

3.2.b   *Experimental procedure*

The experimental technique for confined compression and extension tests is similar to that of von Kármán (1911) and Böker (1915). The conventional triaxial compression apparatus designed by Brace (1964) was used. The confining pressure was first applied; then, axial load was increased in compression tests or decreased in extension tests until fracture occurred. Two of the principal stresses are equal to the confining pressure. Measurements of confining pressure were made by a manganin gage which was calibrated by the Heise Bourdon tube gage. Accuracy of pressure measurements was about 1–2 MPa for pressure less than 100 MPa and about ±1% for higher pressure. Axial loads were measured by a load cell of electric resistance strain gage type outside the pressure vessel, and the contribution of the friction between the ram and the pressure seal was subtracted from the axial load. Accuracy of axial stress measurements was about 2% for stress higher than 100 Mpa. Strain was measured directly by an electric resistance strain gage mounted to the specimen.

3.2.b.1   *Confined compression test*
The precise measurement of compressive strength of a cylindrical rock specimen under fluid-confining pressure was carried out by use of the revised specimen design which was explained in Section 1.1 (Mogi, 1966). The specimen under confining pressure was jacketed by silicone rubber (Fig. 3.5a).

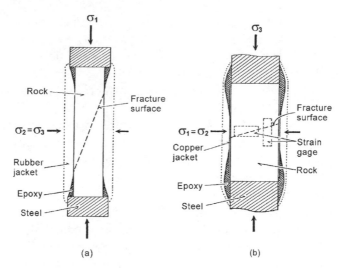

Figure 3.5.   (a) Test specimen for uniaxial and confined compression test. (b) Specimen for confined extension test.

3.2.b.2   *Confined extension test*

Extension tests were also carried out using a new specimen design, as shown in Fig. 3.5(b). The axial stress σ_3 is markedly smaller than the lateral stress $\sigma_1 = \sigma_2$ (confining pressure), and the determination of σ_3 needs various corrections, which are often comparable to total values of the axial stress. The limited accuracy of extension tests is mainly due to this circumstance. In this study, the axial stress at failure was carefully measured using the following techniques.

• *Minimization of error due to sample shape*

For application of σ_3 of wide range including tensile stress, 'dogbone' shaped specimens were used by some investigators (for example, Brace, 1964; Murrell, 1965). The experimental arrangement by Brace (1964) is shown in Fig. 3.6(b). However, the large difference in cross-sectional areas between the central part and the end part of specimen introduced a considerable error, particularly at high confining pressures. The error is mainly caused by the limited accuracy in measurement of confining pressure. Therefore, a dogbone shaped specimen with the very small difference of the cross-sectional areas was used in this experiment, as shown in Fig. 3.6(c). The specimen was connected directly to the rams with epoxy cement. Faults almost always occurred at the central part of the specimen.

• *Correction of the lateral deformation of specimen*

Since axial failure stresses were very small as compared with confining pressures, the change in the cross-sectional area of the specimen by lateral deformation considerably effects a change of the observed axial load. This situation in extension tests is quite different from that in compression tests. The correction of the axial stress ($\Delta\sigma_1$) due to the lateral deformation is

$$\Delta\sigma_1 = (\Delta S/S)P \tag{3.1}$$

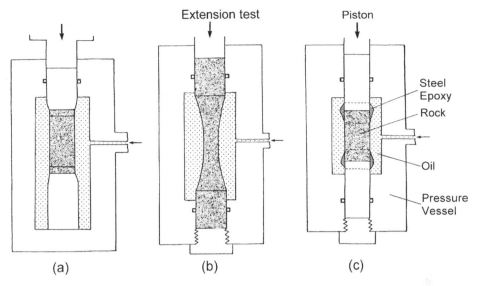

Figure 3.6. Various methods for extension tests under confining pressure ($\sigma_1 = \sigma_2 > \sigma_3$). (a) Conventional method. (b) Method by Brace (1964). (c) Method by Mogi (1967).

where S is the original area of the specimen, ΔS is the decrease in area of the central part of the specimen, and P is the confining pressure. For example, when $P = 500$ MPa and $\Delta S/S = 0.04$, $\Delta \sigma_1$ becomes 20 MPa. This correction is comparable to the values of axial stress at failure. Therefore, the lateral strain in the central part of the specimen at failure should be measured precisely in extension tests. In most previous experiments, however, the correction has been made by using the lateral strain which was *roughly* estimated from the axial strain measured by a dial gage outside the pressure vessel (for example, Böker, 1915; Heard, 1960). In the present experiment, the lateral strain was precisely measured by electric resistance strain gages mounted directly on the specimen (within an accuracy of 0.001), as shown in Fig. 3.5(b).

• *Correction of friction between the ram and the pressure seal*
Because confining pressures were very high, the friction between the ram and the pressure seal made it difficult to measure precisely the axial stress. The frictional effect was determined experimentally by measuring the force required to advance or retract the ram at given hydrostatic pressures. Although reproducibility of the friction measurements was relatively low, corrections were known within an accuracy of about 1 MPa.

• *Correction for jacketing*
Specimens in extension tests were jacketed with copper or steel foil, 0.05 mm thick (Fig. 3.5(b)). The contribution of the jacketing materials to total stresses measured at failure must be taken into account, particularly for steel jacketing. The correction for jacketing was made by assuming the following relation:

$$F = S_r \sigma_r + S_j \sigma_j \tag{3.2}$$

where F is the measured total axial force at failure, S_r and S_j are the sectional areas of the rock specimen and the jacket, respectively, σ_r is the failure stress σ_3 of the rock specimen, and σ_j is the tensile strength of the jacketing material. σ_j was measured under atmospheric conditions. The correction for jacketing was about 1.5 MPa for copper and about 13 MPa for steel. This method of correction was checked by comparing experiments with the two different jacketing materials that gave roughly similar results.

Thus, the accuracy of axial stresses σ_3 with all corrections was probably within 2–3 MPa. Confining pressures $\sigma_1 = \sigma_2$ were accurate to about 1%.

3.2.c *Specimen materials*

Three rocks were used in the experiments: Westerly granite, Dunham dolomite (Block 1), and Solnhofen limestone. Their mechanical homogeneity and isotropy were examined by uniaxial compression tests, as described in Table 3.2. Another rock, an unknown white marble, was also examined, but the isotropy was insufficient for the present purpose. All specimens were taken from a single block of each rock. Compression specimens were circular cylinders 1.6 cm in diameter and 5.0 cm long. Extension specimens were from 2.30 to 2.54 cm in diameter and 5.0 cm long.

3.2.d *Experimental results*

Confined compression and extension tests were made on all three rocks. The reproducibility of strength measurements was examined by uniaxial compression tests for specimens having three orientations at right angles to one another. In the present range of experimental conditions, behavior was brittle in most cases. Stresses at failure then coincided with the maximum stress achieved during an experiment, so that these values are unequivocally determined. In some tests under the higher confining pressures, Solnhofen limestone exhibited ductile behavior, characterized by the deformation curve without any downward breaks in slope after the yield stress. Here the yield stresses were taken as failure stresses. The determination of yield stresses is sometimes uncertain because of a lack of a marked break in the stress-strain curves. Since the curves in this experiment had a marked break at yielding, the stresses were determined with good accuracy. The accuracy is shown in Figs. 3.7–3.9.

3.2.d.1 *Examination of isotropy and homogeneity by uniaxial compression tests*
Uniaxial compressive strengths of the three groups of specimens oriented at right angles to one another were measured for Westerly granite, Dunham dolomite, Solnhofen limestone, and a medium-grained, white marble. The strength values in the three orientations 1, 2, and 3 are shown in Table 3.2. The reproducibility in strength measurements in the same direction was very good, even in the marble. The high reproducibility shows the homogeneous structure of these rocks and the high accuracy of the present measurements. There are some differences of strength in different directions. This directional dependence of strength may be attributed to

Table 3.2. Summary of uniaxial compression tests.

Experiment No.	Orientation		Compressive strength (MPa)	Relative Strength (%)	Anisotropic factor (%)
Westerly granite					
1	1		243		
2	1		252		
Average	1		248	97	
6	2		251		
5	2		253		
Average	2		252	98	
4	3		266		
3	3		273		
Average	3		269	105	
		Total	256 ± 10	100 ± 4	8
Dunham dolomite					
1	1		210		
2	1		214		
Average	1		212	98	
4	2		213		
3	2		215		
Average	2		214	99	
5	3		225		
6	3		226		
Average	3		225	104	
		Total	217 ± 6	100 ± 3	6
Solnhofen limestone					
3	1		288		
4	1		288		
Average	1		288	98	
5	2		298		
6	2		292		
Average	2		295	100	
1	3		300		
2	3		292		
Average	3		296	101	
		Total	293 ± 5	100 ± 2	3
Marble					
1	1		48.9		
3	1		49.6		
2	1		51.0		
Average	1		49.8	89	
6	2		52.5		
4	2		53.5		
5	2		56.7		
Average	2		54.2	97	
7	3		62.0		
8	3		62.8		
9	3		65.7		
Average	3		63.5	114	
		Total	55.8 ± 5.9	100 ± 11	25

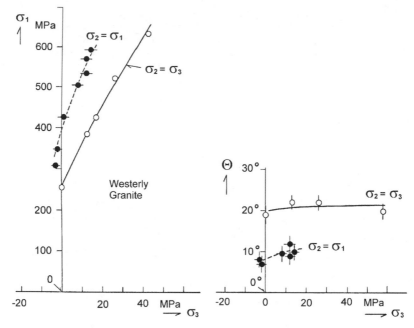

Figure 3.7.   Failure stress (σ_1) and fracture angle (θ) as functions of the minimum principal stress (σ_3) in Westerly granite. Open circle: compression; solid circle: extension; bar: experimental error.

the anisotropic structure. The largest differences for three mutually perpendicular directions are also shown in Table 3.2 in per cent of the average strength value. This difference was less than 8% for the rocks except the marble, for which the difference was very large (about 25%). In the present study, 10 to 50% differences of strength are associated with the influence of the intermediate principal stress. Therefore, Dunham dolomite, Solnhofen limestone, and Westerly granite were suitable enough as test materials, although uncertainties due to their anisotropy should be taken into account in the discussion.

It is clear, however, that highly anisotropic rocks, such as the marble, were unsuitable for the present purpose.

3.2.d.2   *Comparison of confined compression and extension tests*

Data from compression and extension tests are plotted in Figs. 3.7, 3.8 and 3.9, and numerical data are shown in Table 3.3. If the intermediate principal stress is without influence, the curve of σ_1 versus σ_3 for extension should coincide with the curve for compression tests. In Westerly granite and Dunham dolomite, the curves for both tests are nearly parallel, but the extension curve lies well above the compression curve. The differences of the maximum principal stress between the two different tests were 30 to 50% of the value in compression. This discrepancy, which is considerably greater than that obtained by other investigators, is significant enough, even if probable errors are considered. From this fact it is concluded that the intermediate principal

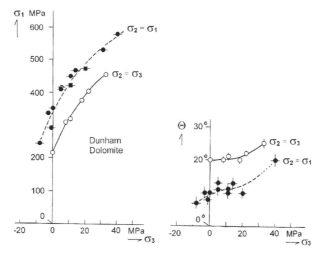

Figure 3.8.   Failure stress (σ_1) and fracture angle (θ) as functions of the minimum principal stress (σ_3) in Dunham dolomite. Open circle: compression; solid circle: extension with copper jacked; solid square: extension with steel jacket.

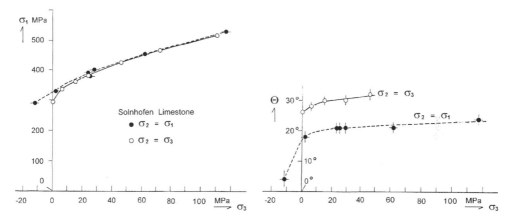

Figure 3.9.   Failure stress (σ_1) and fracture angle (θ) as functions of the minimum principal stress (σ_3) in Solnhofen limestone. Open circle: compression; solid circle: extension.

stress has marked influence on failure conditions in these rocks. The fractures in the present experiment were clearly of shear type at high σ_3 values, but at negative or low positive value of σ_3 they were often curved and irregular. The angle between the fracture surface and the maximum principal stress axis is plotted in Figs. 3.7 to 3.9.

Although measurements of fracture angles contain some uncertainty because of the irregular shapes of fracture surfaces, angles could be determined within ±2°. In these rocks, fracture angles observed in extension tests were 30 to 50% lower than those in compression tests at a given value of the minimum principal stress. This great difference in fracture angles between compression and extension tests is consistent

Table 3.3.   Summary of triaxial compression and extension tests.

Test	σ_1 (MPa)	σ_2 (MPa)	σ_3 (MPa)	θ (°)	Jacketing	Mode of failure
Westerly granite						
Compression	256	0	0	19		Brittle
	385	13	13	22	Rubber	Brittle
	424	17	17	...	Rubber	Brittle
	520	26	26	22	Rubber	Brittle
	632	42	42	...	Rubber	Brittle
	758	58	58	20	Rubber	Brittle
Extension	308	308	−3	8 ?	Copper	Brittle
	348	348	−2	7	Copper	Brittle
	426	426	1	...	Copper	Brittle
	504	504	8	9	Copper	Brittle
	533	533	12	12	Copper	Brittle
	570	570	12	9 ?	Copper	Brittle
	591	591	14	10	Copper	Brittle
Dunham dolomite						
Compression	217	0	0	20		Brittle
	308	8	8	20	Rubber	Brittle
	318	11	11	21	Rubber	Brittle
	375	18	18	20	Rubber	Brittle
	403	22	22	22	Rubber	Brittle
	454	33	33	25	Rubber	Transitional
Extension	243	243	−8	7	Copper	Brittle
	291	291	−0.8	8	Copper	Brittle
	336	336	−3	10	Copper	Brittle
	351	351	0	10	Copper	Brittle
	410	410	5	13	Copper	Brittle
	411	411	5	11	Copper	Brittle
	422	422	11	10	Steel	Brittle
	449	449	11	11	Copper	Brittle
	467	467	14	13	Copper	Brittle
	474	474	20	10	Steel	Brittle
	531	531	31	...	Copper	Brittle
	581	581	40	20	Copper	Brittle
Solnhofen limestone						
Compression	293	0	0	26 ?	...	Brittle
	335	6	6	28	Rubber	Brittle
	360	15	15	30	Rubber	Brittle
	381	24	24	30	Rubber	Brittle
	426	46	46	32 ?	Rubber	Transitional
	467	72	72	...	Rubber	Ductile
	518	111	111	...	Rubber	Ductile
	595	195	195	...	Rubber	Ductile
	709	304	304	...	Rubber	Ductile
Extension	290	290	−12	4	Copper	Brittle
	329	329	2	18	Copper	Brittle
	378	378	25	21	Copper	Brittle
	392	392	24	21	Copper	Brittle
	401	401	28	21	Copper	Brittle
	454	454	62	21	Copper	Brittle
	530	530	117	24	Copper	Brittle

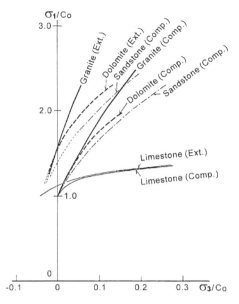

Figure 3.10. Summary of failure conditions in compression and extension tests, including results obtained by Murrell (1965) on Darley Dale sandstone. C_0: uniaxial compressive strength. (Mogi, 1967).

with the value obtained by Böker (1915), Brace (1964), and other investigators. This difference is also attributed to the influence of the intermediate principal stress.

In Solnhofen limestone a marked difference between the failure stress curves for compression and extension tests was not found. The extension curve lies also above the compression curve at low values of σ_3, but both curves roughly agree at higher values of σ_3. The apparent coincidence of the two curves does not always mean that the intermediate principal stress is without influence, because Solnhofen limestone showed ductile behavior at high σ_3. However, it is concluded that the influence of the intermediate principal stress on failure stress is at least not large. This small effect of σ_2 might be connected with the fact that Solnhofen limestone behaves as a ductile material at the higher confining pressure. Marked differences in fracture angles between compression and extension tests were still found. Fracture angles were about 30% lower in extension tests than in compression tests at a given value of the minimum principal stress. Among previous works, Murrell's (1965) experiments on Darly Dale sandstone seemed to be good one. His experimental results are consistent with the above-mentioned result. In Fig. 3.10, observed results including Murrell's one for compression and extension tests are summarized in principal coordinates, σ_1 and σ_3, normalized by dividing by the uniaxial compressive strength (C_0). The following conclusions are obtained:

1. Each curve is slightly concave downward. The slope of these curves at any value of σ_3/C_0 is different for different rock types. The extension curve is nearly parallel

to the compression curve, but the extension curve lies appreciably above the compression curve.

2. The difference of σ_1/C_0 between compression and extension tests at a given value of σ_3/C_0 is proportional to the slope of the compression curve. This result suggests that the σ_2 dependency of the failure stress is proportional to the σ_3 dependency, but smaller.

3. Fracture angles between the maximum principal stress and the shear fracture plane are markedly smaller in extension tests than in compression tests.

These results are inconsistent with current failure criteria, such as the Coulomb, Mohr, Griffith, and modified Griffith criteria, which do not consider the σ_2 effect.

3.3   TRUE TRIAXIAL COMPRESSION EXPERIMENTS

3.3.a   *Introduction*

As mentioned in the preceding section (3.2), it was concluded that the intermediate principal stress (σ_2) had appreciable influence on the fracture strength and the fracture angles of rocks. However, in this discussion, the stress state was a special case in which two principal stresses are equal ($\sigma_2 = \sigma_3$ or $\sigma_1 = \sigma_2$). Therefore, a number of investigators challenged the general (true) triaxial compression experiments in which all three principal stresses are different ($\sigma_1 \geq \sigma_2 \geq \sigma_3$) and attempted to make clear the role of all principal stresses in failure behaviors of rocks. These tests in early stage have included torsion of solid and hollow cylinders under compression, punching under confining pressure, and Brazilian tests under confining pressure (for example, Robertson, 1955; Handin, Higgs and O'Brien, 1960; Jaeger and Hoskins, 1966). The observed discrepancy in different stress states was attributed to the influence of σ_2 (for example, Jaeger and Hoskins, 1966). In these experiments, however, inhomogeneous stress distributions in the rock specimens introduce serious errors in the failure stress calculated by the elasticity theory with a linear stress-strain relation. In addition, there is an unknown effect of the stress gradients.

In principle, triaxial homogeneous stresses, $\sigma_1 \geq \sigma_2 \geq \sigma_3$, seem to be obtained by loading a rectangular parallelepiped across its three pairs of mutually perpendicular surfaces (Fig. 3.1(B)). If stresses are applied through steel end pieces, friction between end pieces and specimens and stress concentration at the ends of specimens could introduce marked errors.

As mentioned before, Handin, Heard and Magouirk (1967) conducted an experiment by subjecting *very thin hollow cylinders* to combined compression and torsion under confining pressure. However, the reproducibility in their torsion test was too low to derive any significant conclusion.

Hojem and Cook (1968) made a new triaxial test, in which lateral stresses (σ_2 and σ_3) were applied by two pairs of thin copper flat jacks. This method seems to achieve homogeneous triaxial stresses, and their result was nearly consistent with the result obtained by the present author (Mogi, 1970a,b). In this method, however, the application of high lateral stresses is difficult because of the limited strength of thin soft copper membrane in the flat-jack. Hojem and Cook presented a few experimental test results in their paper and it seems that they never continued their experiments any further, as mentioned in Section 3.1.

For further study on rock deformation under general stress state, it has been essential to overcome experimental difficulties by achieving the state of true homogeneous triaxial stresses. In 1966, the author designed a new true triaxial compression apparatus, and constructed it in cooperation with Mitsubishi Atomic Power Industries, Inc. Tokyo, Japan. After preliminary experiments, a satisfactory true triaxial compression apparatus was completed in 1968. In 1969, the first reliable experimental results were read at the formal Monthly Meeting of the Earthquake Research Institute, University of Tokyo and at the 1969 Annual Meeting of Seismological Society of Japan (Mogi, 1969, 1970a). By use of this apparatus, deformation and failure behavior of several rocks under the general triaxial stress system have been measured at room temperature and a constant strain rate. In the following sections, this triaxial compression technique is described, and then the experimental results are discussed (Mogi, 1970a, b; 1971a, b; 1972a, b; 1977; 1979; Mogi, Kwaśniewski and Mochizuki,1978; Mogi, Igarashi and Mochizuki, 1978).

3.3.b   *Design of the true triaxial apparatus*

For design of this apparatus, the following conditions should be considered:

1. The three principal stresses can be applied independently.
2. High stresses can be applied.
3. The stress distribution in the specimen is homogeneous (no stress concentration).

To achieve these conditions, a new true triaxial compression apparatus was designed by an important modification of the Kármán-type triaxial testing apparatus as follows.

According to previous experiments, the effect of the minimum principal stress σ_3 on the fracture strength is very large in most brittle rocks. For example, $\Delta\sigma_1 / \Delta\sigma_3$ is about 4 or larger in Dunham dolomite. Therefore, for quantitative discussion of rock fracture, very uniform application of σ_3 and its accurate measurement are necessary. Accordingly, it was decided that σ_3 would be applied by fluid-confining pressure, as in the conventional triaxial test.

On the other hand, in general the effect of the intermediate principal stress σ_2 is much smaller in comparison with that of σ_3. This suggests that σ_2 can be applied through solid end pieces without serious errors under confining pressure. The friction between the surfaces of the specimen and the end pieces can be largely removed by lubrication. According to Mogi (1966), as explained in Section 1.4, the effect of stress concentration at the ends of rock specimen on the triaxial compression strength

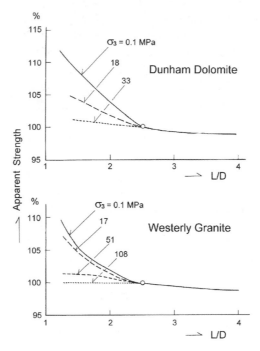

Figure 3.11.   Relation between the apparent fracture strength and the length-diameter ratio (L/D) of rock specimen for different confining pressures. This figure shows that the end-effect of compressed rock specimen decreases markedly with the increase of confining pressure (Mogi, 1966).

greatly decreases with the increase of the confining pressure. The marked decrease of the end effect with the increase of the confining pressure is shown in Fig. 3.11. This effect of the confining pressure plays an important role in achieving a homogeneous triaxial stress state in rock specimens, using this design.

This triaxial compression apparatus was designed from the above-mentioned standpoint, as shown in Figs. 3.12 – 3.15. The most important difference in construction between the conventional triaxial apparatus and the new apparatus are the lateral pistons designed for independent application of σ_2. The pressure vessel is a thick-walled hollow cylinder with diametrical holes for lateral pistons. A confining pressure $p(=\sigma_3)$ of up to 800 MPa can be applied to specimens in this vessel. The axial pistons for the application of σ_1 are connected to a 700 kN Jack (1) and the lateral pistons to a 300 kN Jack (2) (see Fig. 3.12). The axial and lateral loads were measured by load cells inside the pressure vessel.

Another important feature of this design is the ability to keep a rock sample in the same position during the deformation of the sample. This was achieved by a simple mechanical device, as shown schematically in Fig. 3.15. If the upper piston for axial loading is fixed to the pressure vessel, the rock sample must move upward by axial

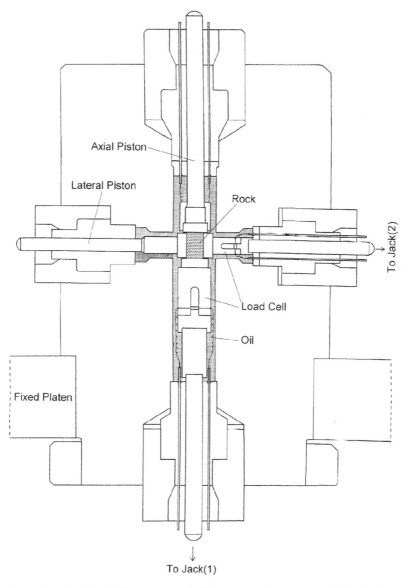

Figure 3.12.   New triaxial cell for the true triaxial compression test in which all three principal stresses are different (vertical section).

loading by Jack (1). On the other hand, the rock sample is fixed in the axial direction by the lateral pistons, and therefore stresses applied to the rock sample might be complex. To solve this problem, frames and pistons for the axial and the lateral loading are set to be freely movable. The frame for the axial piston is kept in the neutral position relative to the pressure vessel by *soft springs* which support the weight of the frame-piston system and so it can move during the deformation of the rock sample under a very small force. The lateral system of piston-Jack and the frame

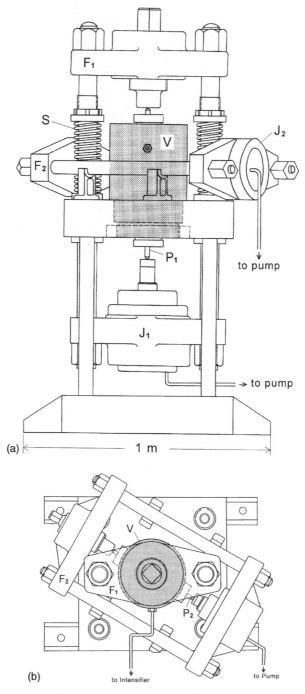

Figure 3.13.   Schematic diagram of the true triaxial compression apparatus. V: Triaxial cell; J_1: jack for axial loading; J_2: jack for lateral loading; F_1: movable frame for axial loading; F_2: movable frame for lateral loading; S: Spring; P_1: axial piston; P_2: lateral piston. (a): front section; (b): horizontal section.

Figure 3.14.   Mogi-type true triaxial machine.

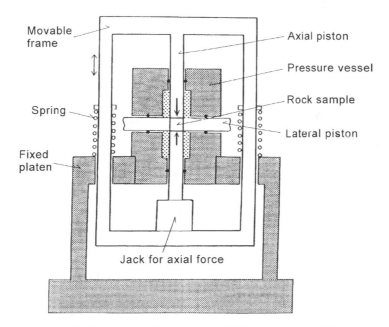

Figure 3.15.   Schematic diagram of the structure of the machine in which a rock sample is kept at the same position during the rock deformation experiment.

can also move freely so the lateral deformation of the rock sample occurs without any unwanted stress application. (In Takahashi and Koide's triaxial apparatus design (1989) mentioned before this problem was not considered, and so the application of the uniform triaxial stresses may be difficult by their machine.)

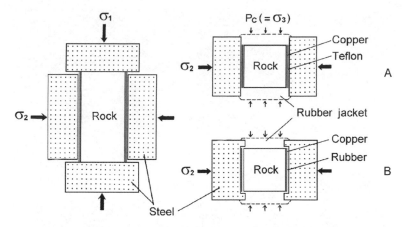

Figure 3.16.   Test specimen for the true triaxial compression test. Left: front section; right: horizontal section in method A and B.

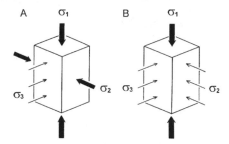

Figure 3.17.   Simplified sketch of method A and B for true triaxial compression. Thick arrow: solid piston; thin arrow: fluidal pressure.

Thus the new triaxial apparatus makes a true triaxial compression test possible, in which each principal stress is homogeneous and controllable independently. This apparatus is not very complex in structure and it was constructed at a limited cost.

3.3.c   *Specimen design and strain measurement*

The test specimen design is schematically shown in Fig. 3.16. The specimen is a rectangular prism 1.5 cm square by 3.0 cm long in many cases. The cylindrical steel end pieces on the top and bottom of the specimen are connected to the specimen with epoxy, and the lateral end pieces are attached to the sides of the specimen using friction-reducing interfaces.

In method A in Figs. 3.16 and 3.17, which was used in most experiments (e.g. Mogi, 1970a,b; 1971a, b; 1977), teflon sheets were used as a lubricant by which the friction between steel and rock surfaces in the σ_2-direction was effectively reduced. To avoid intrusion of teflon into the specimen, the sides of specimen were jacketed by thin copper sheets. The length of the lateral end pieces was slightly shorter than the length of the rock specimen to allow for the axial shortening of the rock under

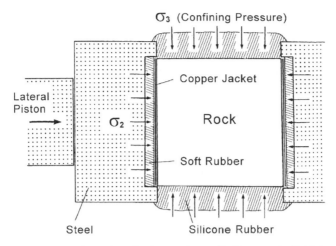

Figure 3.18. Horizontal section of method B. σ_2 is applied by nearly fluidal pressure through thick soft rubber sheets using the special steel end pieces. (Mogi, 1977).

compression. The specimen was jacketed by silicone rubber to prevent intrusion of confining pressure medium (oil) into the specimen.

Although the effect of stress distribution is largely reduced by confining pressure, as mentioned above, the effects of friction and inhomogeneous stress distribution caused by the application of σ_2 by lateral solid pistons may be still apprehended. This problem was checked by the devised method as shown in Figs. 3.16B and 3.18, which corresponds nearly to the case B in Fig. 3.17. In method B, σ_2 was applied by nearly fluidal pressure through thick soft rubber sheets using steel lateral end pieces with a special shape, as shown in Fig. 3.18 (Mogi, 1977). In this case, there is no doubt that σ_2 is applied uniformly.

Figure 3.19 shows the maximum principal stress (σ_1) as a function of the intermediate principal stress (σ_2) obtained by the triaxial apparatus for Yamaguchi marble (Mogi, Igarashi and Mochizuki, 1978), where open circles and closed circles correspond to methods A and B, respectively. As can be seen in this figure, both methods A and B yield almost the same results. This confirms that there is no problem in applying σ_2 by lateral solid pistons. Consequently, method A which is technically simpler to employ, was selected for use in further experiments.

Strains (ε_1, ε_2 and ε_3) in the directions of the three principal stresses σ_1, σ_2 and σ_3 were measured by follows (Mogi, 1970–1977). The axial strain ε_1 and lateral strain ε_2 were measured by electric resistance strain gages mounted on the surface of specimen perpendicular to the σ_3-axis (Fig. 2.2). The axial strain ε_1 was also measured from the displacement of the axial piston, because the electric resistance strain gage is unsuitable for large strains or heterogeneous strains. For this purpose, the true axial deformation of the rock specimen was measured by the method shown in Fig. 2.2 (or 3.87(b)).

The lateral strain ε_3 in the direction of the minimum principal stress σ_3 was measured by the method shown in Fig. 3.87(c). Changes in the width of the specimen in the direction of σ_3 was measured by electric resistance strain gages mounted on the

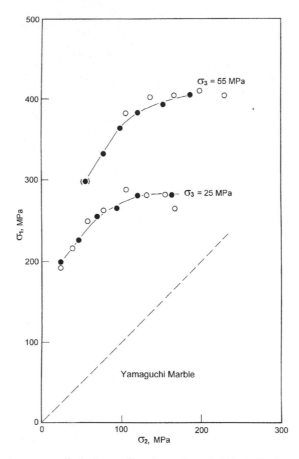

Figure 3.19.   Fracture strength (σ_1) as a function of σ_2. Solid circles by method A and open circles by the method B. (Mogi, Igarashi and Mochizuki, 1978).

curved metal sheets, one end of which was fixed to the surface of the rock specimen and the other end was fixed to the ring-shaped base which is movable in the vertical direction. The deformation meters used in the experiments (Figs. 3.87(b)(c)) were calibrated with a micrometer with a sensitivity of $\pm 10^{-3}$ mm.

3.3.d   *Experimental procedure and rocks studied*

The confining pressure was first applied, and the lateral load was increased to a constant value by Jack (2); then axial load was increased by Jack (1) with a constant strain rate (10^{-4}/sec). Measurement of confining pressure was made both by a manganin gage and a Heise-Bourdon tube gage. The accuracy of pressure measurements was ± 0.1 MPa. Axial and lateral loads were measured by an internal load cell of electric resistance strain gage type. The accuracy of σ_1 and σ_2 measurements was about 2%. Axial strain was measured directly by an electric resistance strain gage mounted to the specimen in most cases, but large strain was measured by the method mentioned above.

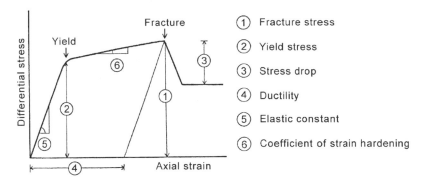

Figure 3.20. Terminology in this book. See text.

In the author's experiments in the 1970s, the following isotropic homogeneous rocks were tested: Dunham dolomite, Solnhofen limestone, Yamaguchi marble, Mizuho trachyte, Orikabe monzonite, Inada granite, and Manazuru andesite. All specimens were taken from a single block of each rock, which is different from the block used in the author's previous experiments. At first, the experimental results on these seven rocks are discussed in detail (e.g. Mogi, 1970b; 1971a, b; 1972a, b; 1977). Other experiments including those carried out on soft, heterogeneous sandstone and an anisotropic rock, and recent experimental results by other investigators are explained later.

3.3.e *Experimental results (1) – Stress-strain curves and fracture stresses*

In Fig. 3.20, the typical stress-strain curve is shown for the terminology used in the following discussions.

1. *Fracture stress* or *fracture strength* is the differential stress $(\sigma_1 - \sigma_3)$ or σ_1 at fracture.
2. *Yield stress* is the deferential stress $(\sigma_1 - \sigma_3)$ or σ_1 at the onset of permanent strain, marked by a sudden break in the stress-strain curve. However, this is indefinite for some rocks, so that the stress at some small permanent strain such as 0.2% is taken in such cases.
3. *Stress drop at fracture.*
4. *Ductility* is qualitatively defined as the ability of the material to undergo large permanent deformation without fracture. There appears to be no universal acceptable numerical measure of ductility. Handin (1966) proposed the total or permanent strain before fracture as the measure of relative ductility. In this book, the permanent strain before fracture is taken as the measure of ductility.
5. *Elastic constant* is obtained from the slope of the curve in the initial linear part, or more strictly from the slope of un- and re-loading curve, as mentioned in Section 2.1.
6. *Coefficient of strain hardening* is obtained from the slope of the curve in the post-yield region.

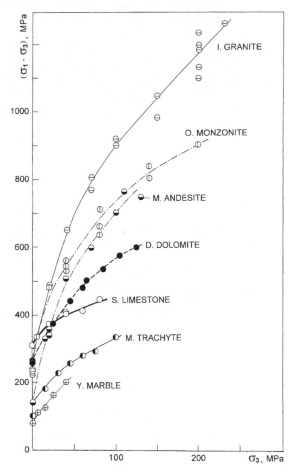

Figure 3.21.   Fracture strength $(\sigma_1 - \sigma_3)$ as functions of the confining pressure of rocks which were tested using the compression apparatus.

7. *Fracture angle* is the angle between the shear fault plane and the axis of the maximum principal stress (σ_1).

The above-mentioned 7 tested rocks show a great variety in mechanical properties. Figure 3.21 shows the fracture strength $(\sigma_1 - \sigma_3)$ as a function of the confining pressure $(p = \sigma_2 = \sigma_3)$ in the conventional triaxial test for these rocks. Figure 3.22 shows the same strength-pressure relations in which stresses are normalized against the uniaxial compressive strength.

3.3.e.1   *Dunham dolomite*

This rock is highly homogeneous and isotropic. A series of stress $(\sigma_1 - \sigma_3)$ – strain (ε) curves obtained under different confining pressures in the conventional triaxial compression test $(\sigma_1 > \sigma_2 = \sigma_3)$ is shown in Fig. 3.23. The ultimate strength and the ductility (ε_n) markedly increase with increasing pressure $(p = \sigma_2 = \sigma_3)$.

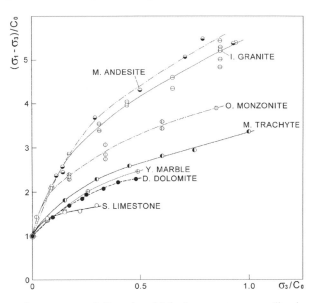

Figure 3.22. Strength-pressure relations in which stresses are normalized against the uniaxial fracture stress of tested rocks.

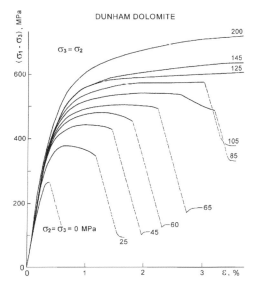

Figure 3.23. Stress $(\sigma_1 - \sigma_3)$ – strain (ε) curves by the conventional triaxial compression test $(\sigma_1 > \sigma_2 = \sigma_3)$ of Dunham dolomite. Numerals for each curve are values of confining pressure in MPa. ε: strain in σ_1-axis.

Each plot in Figs. 3.24(a)–(g) shows a series of stress $(\sigma_1 - \sigma_3)$–strain (ε) curves for different stress σ_2 values, each for the same σ_3 values. This series of deformation curves indicate a marked effect of σ_2 on failure behavior. The ultimate strength under a constant σ_3 increases clearly with increasing σ_2 except for high σ_2 values. In

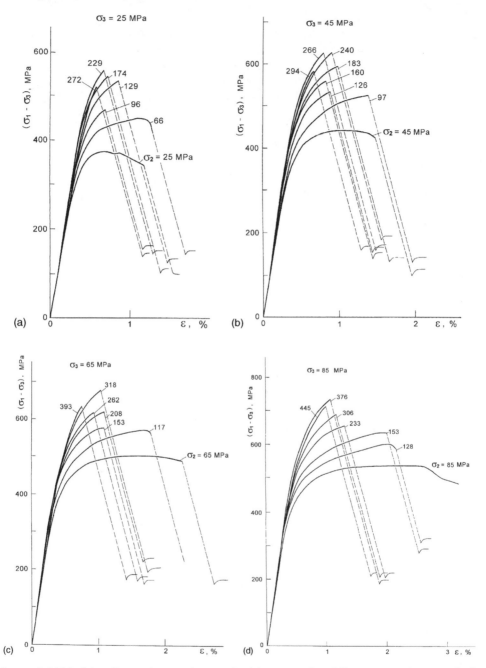

Figure 3.24(a)–(g). Stress $(\sigma_1 - \sigma_3)$ – strain (ε) curves for different σ_2 values, each for the same σ_3, of Dunham dolomite. (a) Dunham dolomite for $\sigma_3 = 25\,\text{MPa}$; (b) Dunham dolomite for $\sigma_3 = 45\,\text{MPa}$; (c) Dunham dolomite for $\sigma_3 = 65\,\text{MPa}$; (d) Dunham dolomite for $\sigma_3 = 85\,\text{MPa}$; (e) Dunham dolomite for $\sigma_3 = 105\,\text{MPa}$; (f) Dunham dolomite for $\sigma_3 = 125\,\text{MPa}$; (g) Dunham dolomite for $\sigma_3 = 145\,\text{MPa}$.

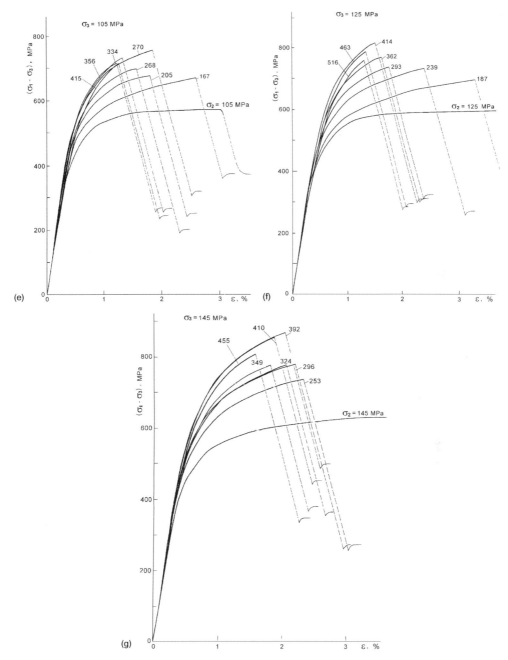

Figure 3.24.   (*Continued*)

these figures, it can be also seen that the ductility (ε_n) considerably decreases with increasing σ_2.

Figure 3.25 shows the maximum principal stress σ_1 at fracture, which is the maximum stress achieved during an experiment in most cases, as a function of the

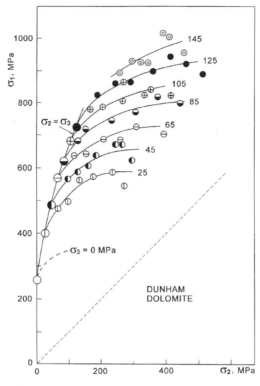

Figure 3.25.   Stress at fracture (σ_1) as functions of σ_2 in Dunham dolomite. Different symbols show the different σ_3, indicated by numerals in MPa. Larger symbols show the case of $\sigma_2 = \sigma_3$.

intermediate principal stress σ_2, for various minimum principal stresses σ_3. The larger symbols on the left side in this graph show the values in the case of $\sigma_2 = \sigma_3$. The σ_1 vs σ_2 curve at $\sigma_3 = 0$ in this figure was obtained from the conventional biaxial test, in which the stress distribution is inhomogeneous to some degree and so the reliability of this curve may be low.

In this figure, the σ_2 effect is large at low σ_2 values and gradually decreases with the increase of σ_2, that is, the σ_1 vs σ_2 curves are concave downward.

Figure 3.26 shows σ_1 as a function of σ_2 and σ_3 in the σ_2–σ_3 plane. The contour lines show equal fracture stress (σ_1) values. From the trend of these contour lines, the ratio of the σ_2 dependency to the σ_3 dependency of the fracture stress (σ_1) is roughly estimated to be about 0.2, except for very low ($\sigma_2 - \sigma_3$) values.

In Fig. 3.27, ε_n, which indicates ductility, is shown as a function of σ_2 and σ_3 in the σ_2–σ_3 plane. The linearity and parallelism of the contour lines showing equal ε_n value are noticeable. From the trend of contour lines, the ratio of the σ_2 dependency to the σ_3 dependency of ductility is estimated to be about -0.3. The trend of contour lines of ductility is nearly perpendicular to the trend of contour lines of fracture stresses. This indicates that the effects of σ_2 and σ_3 on ε_n are the reverse of those on the fracture stress.

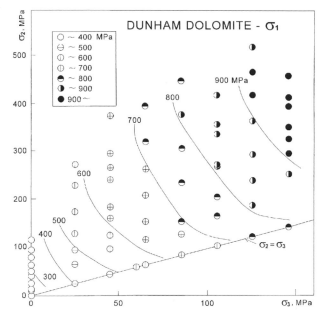

Figure 3.26. σ_1 at fracture as a function of σ_2 and σ_3 in the σ_2–σ_3 plane, in Dunham dolomite.

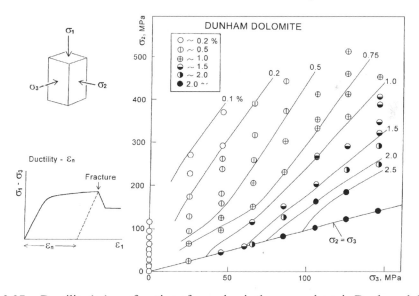

Figure 3.27.   Ductility (ε_n) as a function of σ_2 and σ_3 in the σ_2–σ_3 plane, in Dunham dolomite.

Figure 3.28 shows the stress drop at fracture, as a function of σ_2 and σ_3 in the σ_2–σ_3 plane. The stress drop also increases with increasing σ_2, but it has the maximum values for moderate σ_2 and σ_3. The contour lines showing equal stress drops are somewhat similar to those of ε_n, but more complex.

From the above-mentioned results, it is concluded that the intermediate principal stress σ_2 has an important influence on failure behavior of Dunham dolomite, so a

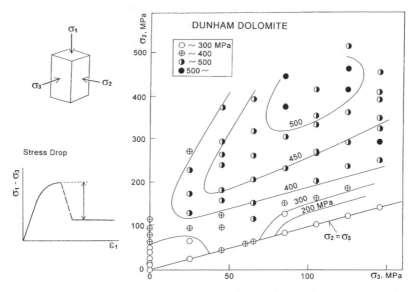

Figure 3.28.   Stress drop at fracture as a function of σ_2 and σ_3 in the σ_2–σ_3 plane, in Dunham dolomite.

discussion of fracture without consideration of σ_2 is not sufficient. From this result, it can be seen that failure behavior under the conventional triaxial compression in which two of the principal stresses are equal is a special one, and the result is not generally applicable in the case of true triaxial stress states.

The numerical data of this experiment are shown in Table 3.4.

3.3.e.2   *Solnhofen limestone*

The deformation and fracture of this rock, which is extremely homogeneous and isotropic, has been studied by many investigators (e.g., Robertson, 1955; Heard, 1960; Brace, 1964; Handin et al., 1967; Mogi, 1967).

Figure 3.29 shows a series of stress-strain curves obtained from the conventional triaxial test ($\sigma_1 > \sigma_2 = \sigma_3$). The ductility greatly increases with the confining pressure and no considerable stress drop was observed within 4 percent strain under the confining pressure of 80 MPa and higher. The strength did not appreciably increase with the confining pressure, except for very low pressure. It is particularly noticeable that the yield stress is nearly independent of the confining pressure. These features are consistent with those described before.

Figure 3.30 shows series of stress-strain curves for different σ_2 values under two constant σ_3 values ($\sigma_3 = 120$ MPa and 70 MPa), obtained by the true triaxial compression test. In the case of $\sigma_3 = 120$ MPa, no fracture was observed within 4 percent strain for different σ_2 values, and the yield stress increases slightly with increasing σ_2. In the case of $\sigma_3 = 70$ MPa, stress-strain curves changed from the ductile type to the brittle type with increase of σ_2 values.

Each figure in Fig. 3.31 (a)–(f) shows series of stress-strain curves for different σ_2 values, each under constant σ_3 values. These series of deformation curves also

Table 3.4.   Summary of true triaxial compression tests of Dunham dolomite (block 2): Fracture.

Experiment No.	σ_1 (MPa)	σ_2 (MPa)	σ_3 (MPa)	$\dfrac{\sigma_1 + \sigma_3}{2}$ (MPa)	τ_{oct} (MPa)	ε_n %	Stress drop (MPa)
154	(265)	0	0	133	125		
162	(258)	0	0	129	122		
156	400	25	25	213	177	0.85	255
128	475	66	25	250	203	0.79	291
139	495	96	25	260	207	0.24	313
118	560	129	25	293	232	0.38	411
137	571	174	25	298	230	0.24	440
129	586	229	25	306	231	0.19	415
160	545	272	25	285	212	0.11	382
103	487	45	45	266	208	1.09	328
107	570	97	45	308	236	0.90	390
101	576	126	45	311	234	0.42	375
104	606	160	45	326	242	0.40	415
109	639	183	45	342	254	0.48	462
102	670	240	45	358	261	0.38	444
144	670	266	45	358	259	0.32	466
111	622	294	45	334	236	0.20	415
113	540	60	60	300	226	1.45	328
124	568	65	65	317	237	1.83	328
140	638	117	65	352	259	1.30	
114	644	153	65	355	255	0.63	415
112	687	208	65	376	266	0.58	400
125	685	262	65	375	259	0.43	444
115	746	318	65	406	281	0.44	484
116	701	393	65	383	260	0.24	462
145	620	85	85	353	252	2.40	152
153	684	128	85	385	273	1.70	298
136	719	153	85	402	284	1.46	360
143	744	233	85	415	282	0.76	451
135	773	306	85	429	287	0.60	491
147	818	376	85	452	301	0.50	546
155	798	445	85	442	291	0.38	502
131	682	105	105	394	272	2.56	178
120	778	167	105	442	304	2.05	310
121	786	205	105	446	300	1.22	437
134	805	268	105	455	299	1.02	502
141	863	270	105	484	325	1.25	448
108	824	334	105	465	300	0.71	455
142	840	356	105	473	305	0.73	470
138	822	415	105	464	294	0.62	477

(*continued*)

Table 3.4.   (*Continued*)

Experiment No.	σ_1 (MPa)	σ_2 (MPa)	σ_3 (MPa)	$\dfrac{\sigma_1 + \sigma_3}{2}$ (MPa)	τ_{oct} (MPa)	ε_n %	Stress drop (MPa)
172	725	125	125	425	283		
158	824	187	125	475	316	2.73	351
174	860	239	125	493	323	1.76	481
119	863	293	125	494	316	1.12	426
176	897	362	125	511	323	0.95	475
106	941	414	125	533	338	0.82	519
171	918	463	125	522	325	0.62	507
105	886	516	125	506	311	0.62	485
159	*	145	145				
130	883	253	145	514	325	1.72	480
173	927	296	145	536	339	1.58	524
148	923	324	145	534	333	1.40	424
146	922	349	145	534	329	1.13	408
133	1015	392	145	580	366	1.34	383
157	1002	410	145	574	358	1.18	414
123	952	455	145	549	332	0.88	470
152	*	200	200				

* Ultimate stress was not determined in the test because of high ductility.

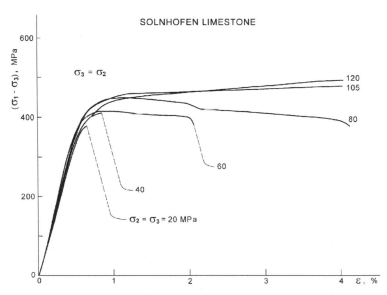

Figure 3.29.   Stress ($\sigma_1 - \sigma_3$) – strain (ε) curves of Solnhofen limestone by the conventional triaxial tests ($\sigma_1 > \sigma_2 = \sigma_3$) of Solnhofen limestone. Numerals for each curve are confining pressure in MPa. ε : axial strain.

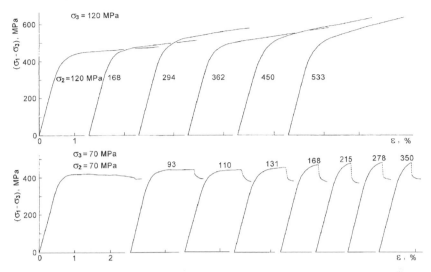

Figure 3.30.   Series of stress ($\sigma_1 - \sigma_3$) – strain (ε) curves for different σ_2 values under two constant σ_3 values (120 MPa and 70 MPa) in Solnhofen limestone.

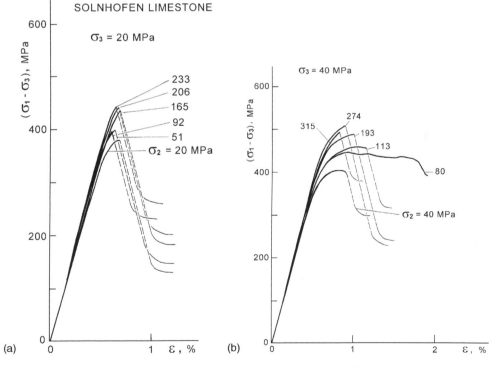

Figure 3.31(a)–(f).   Stress ($\sigma_1 - \sigma_3$)–axial strain (ε) curves for different σ_2 values, each for the same σ_3, of Solnhofen limestone. (a) Solnhofen limestone for $\sigma_3 = 20$ MPa; (b) Solnhofen limestone for $\sigma_3 = 40$ MPa; (c) Solnhofen limestone for $\sigma_3 = 60$ MPa; (d) Solnhofen limestone for $\sigma_3 = 80$ MPa; (e) Solnhofen limestone for $\sigma_3 = 105$ MPa; (f) Solnhofen limestone for $\sigma_3 = 120$ MPa.

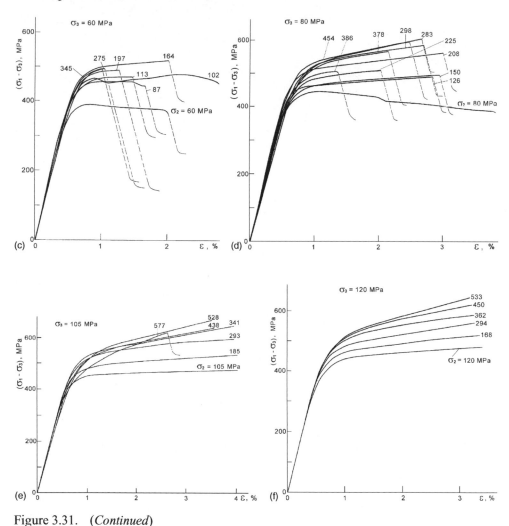

Figure 3.31.   (*Continued*)

indicate a marked effect of σ_2, nearly similar to that in Dunham dolomite, mentioned above.

In Fig. 3.32, the maximum principal stress σ_1 at fracture is plotted as a function of the intermediate principal stress σ_2. The general feature in the σ_1 vs σ_2 relation of this rock is similar to that of dolomite, but the σ_2 effect ($\Delta\sigma_1/\Delta\sigma_2$) is appreciably smaller than that observed in the dolomite, as the effect of the confining pressure ($p = \sigma_2 = \sigma_3$) is smaller. This result is consistent with the previous conclusion derived from the comparison between the conventional triaxial compression and the extension tests (Mogi, 1967).

Figure 3.33 shows the ductility ε_n as a function of σ_2 and σ_3. The contour lines showing equal ε_n values are nearly linear and parallel to each other, as was also the case in the dolomite. ε_n increases markedly with increasing σ_3, but decreases with

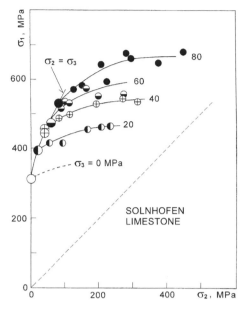

Figure 3.32.   Stress at fracture (σ_1) as functions of σ_2 in Solnhofen limestone. Different symbols show the different σ_3, indicated by numerals in MPa. Larger symbols show the case of $\sigma_2 = \sigma_3$.

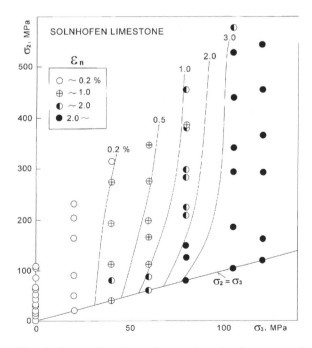

Figure 3.33.   Ductility (ε_n) as a function of σ_2 and σ_3 in the σ_2–σ_3 plane, in Solnhofen limestone.

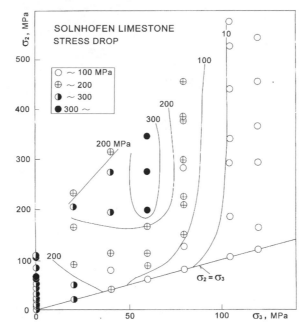

Figure 3.34.   Stress drop at fracture as a function of σ_2 and σ_3 in the σ_2–σ_3 plane, in Solnhofen limestone.

increasing σ_2. The trend of contour lines of ε_n of Solnhofen limestone is essentially similar to that of Dunham dolomite but the contour lines of the limestone cross the σ_2 axis at very small angles. From the trend of curves, the ratio of the σ_2 dependency of ε_n to the σ_3 dependency is roughly estimated to be -0.05, which is markedly smaller than the value obtained for the dolomite.

The stress drop is shown as a function of σ_2 and σ_3 in Fig. 3.34. Although the limiting value of σ_1 and ε_n change monotonically on the σ_2–σ_3 plane, as mentioned above, the stress drop has a maximum value in a certain region. The trend of the axis of the maximum stress drop region is nearly parallel to the σ_2 axis. That is, the σ_3 dependency is much greater than the σ_2 dependency.

In conclusion, the effect of the triaxial stress state in Solnhofen limestone is similar to that in Dunham dolomite. Quantitatively, however, both the σ_2 and σ_3 dependencies of fracture stress are appreciably lower in the limestone than in the dolomite. In particular, the influence of σ_2 on the ductility and the stress drop is considerably lower in the limestone than in the dolomite.

The numerical data of this experiment are shown in Table 3.5.

3.3.e.3   *Yamaguchi marble*

This rock is a medium-grained calcite marble quarried at Mine, Yamaguchi prefecture, Japan. Figure 3.35 shows a series of stress-strain curves under different confining pressures obtained from conventional triaxial test ($\sigma_1 > \sigma_2 = \sigma_3$). This rock shows ductile behavior at confining pressure of 40 MPa and higher. In this figure,

Table 3.5. Summary of true triaxial compression tests of Solnhofen limestone: Fracture

Experiment No.	σ_1 (MPa)	σ_2 (MPa)	σ_3 (MPa)	$\dfrac{\sigma_1 + \sigma_3}{2}$ (MPa)	τ_{oct} (MPa)	ε_n %	Stress drop (MPa)
200	310	0	0	155	146		
218	397	20	20	209	178	0.19	236
219	417	51	20	219	180	0.09	226
220	413	92	20	217	171	0.05	137
217	453	165	20	237	180	0.07	151
221	460	206	20	240	180	0.06	219
224	465	233	20	243	182	0.05	164
210	449	40	40	245	193	0.34	187
203	446	40	40	243	191	0.33	100
206	486	80	40	263	201	1.18	59
205	499	113	40	270	201	0.54	134
201	530	193	40	285	205	0.34	234
204	547	274	40	294	207	0.22	253
209	535	315	40	288	203	0.15	111
249	473	60	60	267	195	1.54	98
256	517	87	60	289	209	1.08	121
247	537	102	60	299	216		
254	530	113	60	295	210	0.84	154
243	576	164	60	318	223	1.40	120
251	550	197	60	305	206	0.64	328
246	553	275	60	307	202	0.40	321
252	557	345	60	309	204	0.35	334
232	528	80	80	304	211	3.44	
240	572	126	80	326	222	2.23	92
244	577	150	80	329	220	2.28	104
236	647	208	80	364	243	2.22	104
233	591	225	80	336	215	1.32	145
234	677	283	80	379	248	1.90	76
238	665	298	80	373	241	1.92	144
231	650	378	80	365	233	1.32	138
222	680	454	80	380	247	1.61	166
230	*	105	105				
235	*	185	105				
225	711	293	105	408	253	3.35	61
226	*	341	105				
229	*	439	105				
227	*	528	105				
228	727	577	105	416	265	1.78	78
211	*	120	120				
216	*						
215	*	294	120				
212	*	362	120				
213	*	450	120				
214	*	533	120				

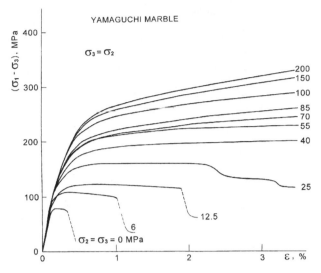

Figure 3.35. Stress $(\sigma_1-\sigma_3)$ – strain (ε) curves by the conventional compression tests $(\sigma_1 > \sigma_2 = \sigma_3)$ of Yamaguchi marble. Numerals for each curve are confining pressure in MPa. ε : axial strain.

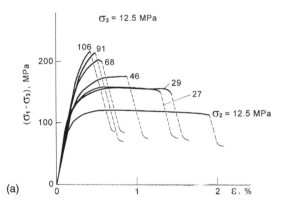

(a)

Figure 3.36(a)–(g). Stress $(\sigma_1-\sigma_3)$ – strain (ε) curves for different σ_2 values, each for the same σ_3, of Yamaguchi marble. (a) Yamaguchi marble for $\sigma_3 = 12.5$ MPa; (b) Yamaguchi marble for $\sigma_3 = 25$ MPa; (c) Yamaguchi marble for $\sigma_3 = 40$ MPa; (d) Yamaguchi marble for $\sigma_3 = 55$ MPa; (e) Yamaguchi marble for $\sigma_3 = 70$ MPa; (f) Yamaguchi marble for $\sigma_3 = 85$ MPa; (g) Yamaguchi marble for $\sigma_3 = 100$ MPa.

the marked increase in ultimate strength or yield strength with confining pressure is noticeable.

Each plot in Figs. 3.36(a)–(g) shows a series of stress-strain curves for different σ_2 values and the same σ_3 values. The increase in strength and the decrease in ductility with increasing σ_2 can be very clearly seen. For the confining pressure $p(=\sigma_3)$ of 100 MPa and higher, no appreciable stress drop was observed even for very

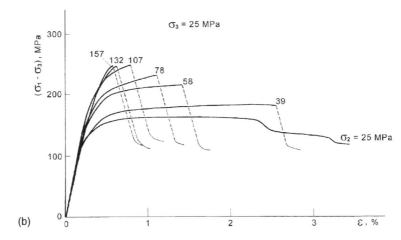

Figure 3.36. (*Continued*)

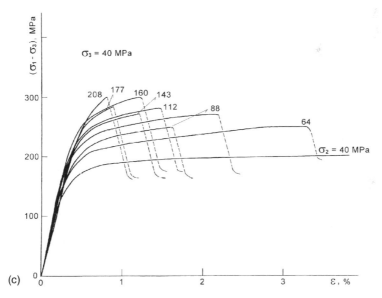

(c)

Figure 3.36. (*Continued*)

high σ_2 values (Fig. 3.36(g)). The influence of the lateral stress on the deformation curve is essentially similar to that observed in the two carbonate rocks mentioned above. The dilatancy of this rock under true triaxial compression is discussed in a later section.

In Fig. 3.37, the maximum principal stress σ_1 at fracture is plotted as a function of σ_2. Different symbols indicate different σ_3 values and larger symbols on the left side in the graph indicate the case where $\sigma_2 = \sigma_3$. The σ_1 values at fracture are greatly influenced by σ_3 and σ_2, and the σ_1 vs σ_2 curves are appreciably concave downward.

Figure 3.38 shows σ_1 values in the σ_2–σ_3 plane. There is a lack of data for a low $(\sigma_2 - \sigma_3)$ region, because no fracture occurred in this region due to higher ductility.

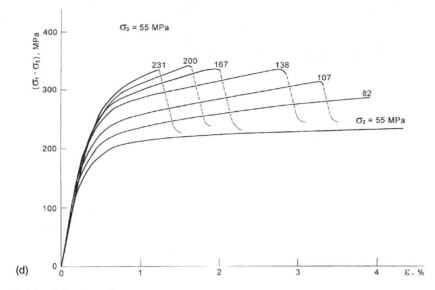

(d)

Figure 3.36.   (*Continued*)

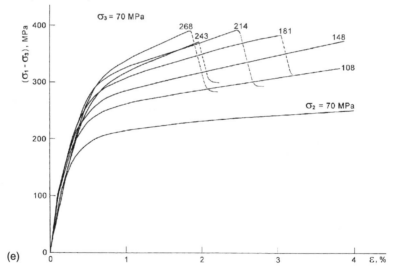

(e)

Figure 3.36.   (*Continued*)

Figure 3.39 shows the ductility (ε_n) as a function of σ_2 and σ_3. The contour lines showing equal ε_n values are nearly linear and parallel to each other. From the trend of contour lines, the ratio of the σ_2 dependency to the σ_3 dependency is roughly estimated to be -0.25. In this rock, the marked ductile behavior appears at lower confining pressure than in the dolomite.

The stress drop is shown as a function of σ_2 and σ_3 in Fig. 3.40. The pattern of the contour lines representing equal stress drops is similar to that of ε_n, although it is more complex. In this case, the maximum values of stress drop locate in the region of higher σ_2 with the contour lines nearly parallel to the contour lines of ε_n.

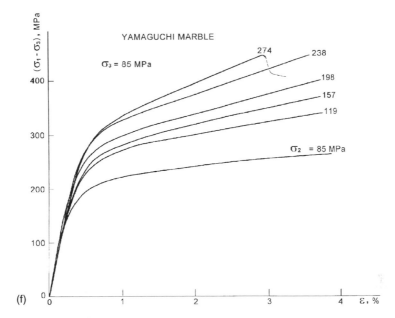

Figure 3.36.   (*Continued*)

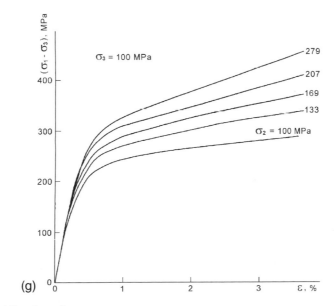

Figure 3.36.   (*Continued*)

By all accounts, the effect of the intermediate principal stress σ_2 on the fracture behavior in Yamaguchi marble is nearly similar to that in Dunham dolomite and larger than that in Solnhofen limestone. The pressure-dependency of yield stress in this rock will be discussed in a later section.

The numerical data of this experiment are shown in Table 3.6.

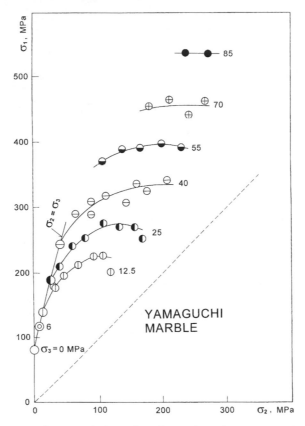

Figure 3.37.   Stress at fracture (σ_1) as functions of σ_2 in Yamaguchi marble. Different symbols show the different σ_3, indicated by numerals in MPa. Larger symbols show the case of $\sigma_2 = \sigma_3$.

3.3.e.4   *Mizuho trachyte*

This rock, quarried at Tomisaki, Gunma prefecture, Japan, is a moderately porous and fairly homogeneous silicate rock. Figure 3.41 shows a series of stress ($\sigma_1 - \sigma_3$)–strain (ε) curves under different confining pressures in the conventional triaxial test ($\sigma_1 > \sigma_2 = \sigma_3$). The strength appreciably increases with confining pressure. The brittleness manifested by the amount of stress drop at fracture decreases with confining pressure and the deformation curve at 100 MPa shows the typical behavior of the ductile rock. Figs. 3.42 (a) and (b) show a series of stress ($\sigma_1 - \sigma_3$)–strain (ε) curves for various σ_2 values under the two different σ_3 values (45 MPa and 75 MPa). The curves change systematically with increase of σ_2.

Each plot in Figs. 3.43 (a)–(d) shows a series of stress-strain curves for different σ_2 values under constant σ_3 values. The ultimate strength increases appreciably with σ_2, and the stress drop phenomenon is more noticeable for higher σ_2 values, as was also seen in the carbonate rocks mentioned above. However, the deformation curves are somewhat different from those of carbonate rocks. When the differential stress

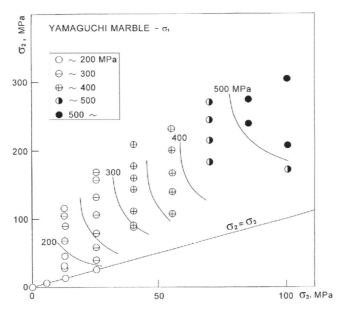

Figure 3.38. σ_1 at fracture as a function of σ_2 and σ_3 in the σ_2–σ_3 plane, in Yamaguchi marble.

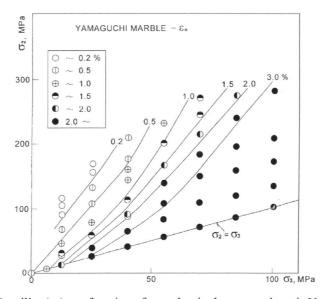

Figure 3.39.   Ductility (ε_n) as a function of σ_2 and σ_3 in the σ_2–σ_3 plane, in Yamaguchi marble.

($\sigma_1 - \sigma_3$) increases in a deformation process, its maximum value appears within a small permanent strain, that is less than 1% and a sudden or gradual decrease in stress follows. This stress decrease is probably caused by cataclastic flow. Therefore, the before-mentioned definition of ductility as the ε_n value is not applicable in this silicate rock. The degree of brittleness or ductility in such cases may be shown by some different measures discussed before (cf. Mogi, 1965).

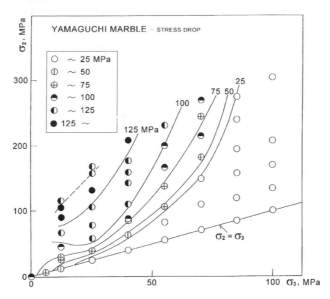

Figure 3.40.   Stress drop at fracture as a function of σ_2 and σ_3 in the σ_2–σ_3 plane, in Yamaguchi marble.

Figure 3.44 shows the limiting values of the maximum principal stress σ_1, which is the maximum stress achieved during an experiment, as a function of the intermediate principal stress σ_2. Different symbols show different σ_3 values and larger symbols show the case of $\sigma_2 = \sigma_3$. The σ_1 vs σ_2 curves are similar to those in carbonate rocks. Figure 3.45 shows the ultimate stress (σ_1) during the fracture process in the σ_2–σ_3 plane, and Fig. 3.46 shows the stress drop as a function of σ_2 and σ_3. The patterns of contour lines in these figures are nearly similar to those of Solnhofen limestone.

Thus, there is no any essential difference in the effect of the intermediate principal stress σ_2 on failure behavior between this silicate rock and the above-mentioned carbonate rocks.

The numerical data of this experiment are shown in Table 3.11.

3.3.e.5   *Manazuru andesite*

This andesite quarried at Manazuru, Kanagawa prefecture, Japan, is a compact, light gray rock. Figure 3.47 shows a series of stress ($\sigma_1 - \sigma_3$)–strain (ε) curves under different confining pressures as obtained from conventional triaxial test ($\sigma_1 > \sigma_2 = \sigma_3$). These are deformation curves typical of brittle materials. Appreciable differences in slopes of the stress-strain curves indicate that this rock is somewhat inhomogeneous.

Figure 3.48 shows the maximum principal stress σ_1 at fracture as a function of the intermediate principal stress σ_2. Different symbols correspond to different σ_3 values and larger symbols show the case where $\sigma_2 = \sigma_3$. The σ_2 dependency of fracture strength is also considerable in this rock, although it is much smaller than

Table 3.6. Summary of true triaxial compression tests of Yamaguchi marble: Fracture.

Experiment No.	σ_1 (MPa)	σ_2 (MPa)	σ_3 (MPa)	$\dfrac{\sigma_1 + \sigma_3}{2}$ (MPa)	τ_{oct} (MPa)	ε_n (%)	Stress drop (MPa)
152	82	0	0	41	39	0.23	
153	118	6	6	62	53	0.90	61
145	140	12.5	12.5	76	60	1.75	48
148	179	26	12.5	96	75	1.10	68
144	177	28	12.5	95	74	1.22	74
150	196	45	12.5	104	80	0.62	98
142	213	67	12.5	113	84	0.27	116
146	225	90	12.5	119	88	0.18	144
147	228	105	12.5	120	86	0.14	134
151	200	115	12.5	106	77	0.06	106
105	189	25	25	107	77		
104	209	39	25	117	84	2.30	65
107	240	58	25	133	95	1.10	106
101	252	78	25	139	97	0.76	108
102	275	107	25	150	104	0.43	124
103	268	132	25	147	99	0.23	126
108	268	157	25	147	99	0.15	122
109	250	168	25	138	93	0.07	111
110	243	40	40	142	96		
111	290	64	40	165	113	3.00	45
112	288	88	40	164	107	1.23	75
114	309	88	40	175	117	1.82	93
113	319	112	40	180	118	1.08	103
115	307	143	40	174	110	0.84	103
118	336	160	40	188	122	0.83	116
116	321	177	40	181	115	0.40	118
117	341	208	40	191	123	0.28	138
119	*	55	55				
120	*	82	55				
121	369	107	55	212	137	2.87	51
122	388	138	55	222	142	2.43	74
123	390	167	55	223	139	1.54	90
124	396	200	55	226	140	1.13	99
125	390	231	55	223	137	0.70	105
154	*	70	70				
141	*	108	70				
136	*	148	70				
139	454	181	70	262	161	2.55	66
143	464	214	70	267	163	1.90	82
140	440	243	70	255	151	1.37	64
138	462	268	70	266	160	1.22	95
155	*	85	85				

(Continued)

Table 3.6.   (*Continued*)

Experiment No.	σ_1 (MPa)	σ_2 (MPa)	σ_3 (MPa)	$\dfrac{\sigma_1 + \sigma_3}{2}$ (MPa)	τ_{oct} (MPa)	ε_n (%)	Stress drop (MPa)
137	*	119	85				
132	*	157	85				
135	*	194	85				
133	535	238	85	310	187		
134	537	274	85	311	185	(1.93)	(23)
129	*	100	100				
128	*	133	100				
130	*	169	100				
126	*	207	100				
127	*	279	100				
159	*	150	150				
160	*	150	150				
164	*	200	200				
165	*	200	200				

the σ_3 dependency. For very large σ_2 values, unexpected lower strength values were sometimes observed.

The numerical data of this experiment are shown in Table 3.8.

3.3.e.6   *Inada granite*
This biotite granite quarried at Inada, Ibaraki prefecture, Japan, is medium-grained and light gray in color. Figure 3.49 shows a series of stress (σ_1) vs strain (ε) curves under different confining pressures as obtained from conventional triaxial test ($\sigma_1 > \sigma_2 = \sigma_3$). Figure 3.50 shows a series of stress ($\sigma_1 - \sigma_3$)–strain (ε) curves for different σ_2 values under a constant σ_3 value ($\sigma_3 = 150$ MPa). The effects of σ_2 and σ_3 on fracture strength are also noticeable. These stress-strain curves are typical of very brittle materials and a slight decrease in the ductility (ε_n) with increasing σ_2 can also be observed. The dilatancy under the true triaxial stress is discussed in a later section.

Figure 3.51 shows the maximum principal stress σ_1 at fracture as a function of the intermediate principal stress σ_2. Different symbols correspond to different σ_3 values and larger symbols show the case where $\sigma_2 = \sigma_3$. Although data somewhat scattered, it may be said that the effects of σ_2 and σ_3 on σ_1 at fracture in this rock are nearly similar to those in the other rocks mentioned above.

The numerical data of this experiment are shown in Table 3.9.

3.3.e.7   *Orikabe monzonite*
This rock was also brittle in the present triaxial compression test. Figure 3.52 shows the maximum principal stress σ_1 at fracture as a function of the intermediate principal stress σ_2. Different symbols show different σ_3 values and larger symbols show the

Table 3.7.   Summary of true triaxial compression tests of Mizuho trachyte: Fracture.

Experiment No.	σ_1 (MPa)	σ_2 (MPa)	σ_3 (MPa)	$\dfrac{\sigma_1 + \sigma_3}{2}$ (MPa)	τ_{oct} (MPa)	ε_n %	Stress drop (MPa)
304	100	0	0	50	47		
327	196	15	15	106	85	0.25	71
332	259	30	30	145	108	0.24	91
307	302	45	45	174	121		24
309	314	58	45	179	124	0.42	44
305	327	67	45	186	128	0.23	60
303	341	90	45	193	130	0.22	98
302	350	138	45	197	128	0.23	95
301	359	204	45	202	128	0.12	104
306	368	281	45	206	136	0.10	95
308	353	323	45	199	139	0.05	71
319	341	60	60	200	132		18
320	353	83	60	206	133		9
321	386	133	60	223	140	0.59	69
318	401	186	60	230	141	0.71	60
315	403	212	60	231	140	0.31	60
317	401	254	60	230	140	0.25	66
322	381	306	60		137	0.21	47
326	368	75	75	221	138		20
329	405	108	75	240	148	0.82	35
325	415	147	75	245	146	0.75	29
330	438	210	75	256	150	0.45	53
323	440	279	75	257	149	0.31	40
324	430	318	75	252	148	0.24	38
331	452	363	75	263	161	0.22	35
312	437	100	100	268	159		4
313	463	126	100	281	165		
314	493	171	100	296	171		18
310	497	256	100	298	163		2
311	522	354	100	311	183		
316	510	384	100	305	171		

case where $\sigma_2 = \sigma_3$. The effects of σ_2 and σ_3 on fracture strength are also nearly similar to those observed in the other rocks and the σ_2 effect seems to be higher for low σ_2 values.

The numerical data of this experiment are shown in Table 3.10.

3.3.e.8 *Summary*

As mentioned above, by employing the new triaxial compression technique, stress $(\sigma_1 - \sigma_3)$–strain (ε) relations for three carbonate rocks and four silicate rocks under

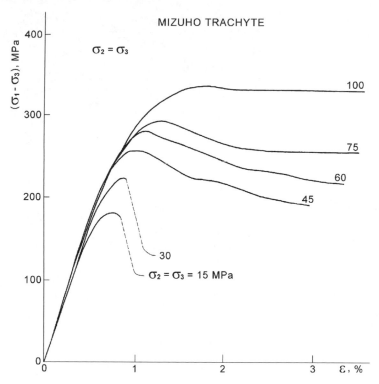

Figure 3.41.   Stress $(\sigma_1 - \sigma_3)$ – strain (ε) curves by the conventional triaxial compression tests $(\sigma_1 > \sigma_2 = \sigma_3)$ of Mizuho trachyte. Numerals for each curve are confining pressure in MPa. ε : axial strain.

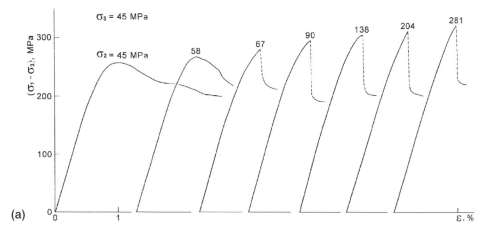

Figure 3.42.   Series of stress $(\sigma_1 - \sigma_3)$ – strain (ε) curves for different σ_2 values under constant σ_3 values in Mizuho trachyte. (a) Mizuho trachyte $\sigma_3 = 45$ MPa; (b) Mizuho trachyte $\sigma_3 = 75$ MPa.

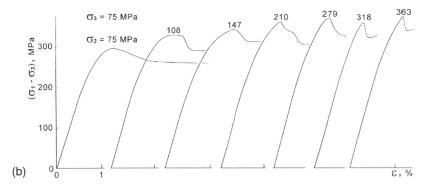

(b)

Figure 3.42. (*Continued*)

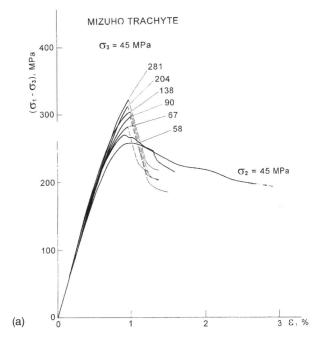

(a)

Figure 3.43(a)–(d). Stress ($\sigma_1 - \sigma_3$) – strain (ε) curves for different σ_2 values, each for the same σ_3, of Mizuho trachyte. (a) Mizuho trachyte $\sigma_3 = 45\,\mathrm{MPa}$; (b) Mizuho trachyte $\sigma_3 = 60\,\mathrm{MPa}$; (c) Mizuho trachyte $\sigma_3 = 75\,\mathrm{MPa}$; (d) Mizuho trachyte $\sigma_3 = 100\,\mathrm{MPa}$.

true triaxial compression conditions ($\sigma_1 \geq \sigma_2 \geq \sigma_3$) were obtained, and the effect of combined stresses, particularly the σ_2 effect on the fracture strength and ductility of these rocks was clearly observed. Although these rocks show different failure behavior, the following general conclusions can be formulated:

(1) The fracture strength increases not only with the increase of σ_3, but also with increase of σ_2, except for very high σ_2 values. The σ_2 dependency is clearly smaller than that of σ_3, but the σ_2 effect is also considerable. In addition, the

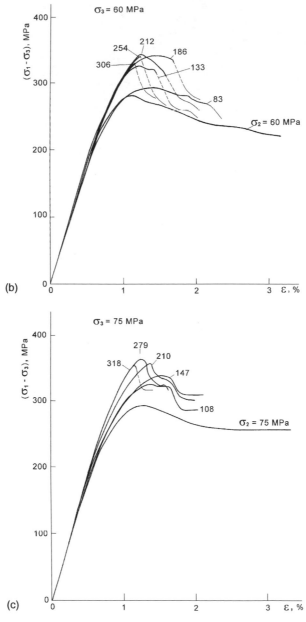

Figure 3.43. (*Continued*)

effects of σ_2 and σ_3 seem to be nearly proportional to each other. The σ_1 vs σ_2 curves are concave downward.

(2) The ductility increases with increasing σ_3, but decreases markedly with increasing σ_2. This effect was observed mainly for carbonate rocks, but it was found even in very brittle silicate rocks, although it was not very pronounced.

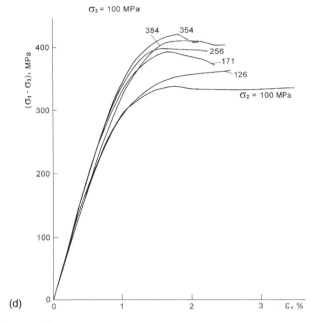

(d)

Figure 3.43. (*Continued*)

(3) Stress drop at fracture generally decreases with increasing σ_3, but increases with increasing σ_2. In this case, however, it has the maximum values for some higher σ_2 values.

(4) The conventional triaxial compression test, in which two of the principal stresses are equal, is a specific one and its results cannot be applied to the general case of the true triaxial state of stresses.

(5) In earlier studies, it was widely stated that the ductility increases with the mean pressure. However, this statement is not correct, because the influences of σ_2 and σ_3 on the ductility are opposite of each other. The effect of the *mean pressure* on failure behavior is complex.

This summary of experimental results is limited mainly to subjects related to fracture, and yielding will be discussed in the next section.

3.3.f   *Experimental results (2) – Yield stresses*

As can be seen in stress-strain curves in the preceding section, the deformation curves in the three carbonate rocks (Dunham dolomite, Solnhofen limestone, Yamaguchi marble) and a silicate rock (Mizuho trachyte) are linear at the initial stage and then abruptly or gradually bend downward. When the slope of curve changes abruptly, the yield stress (σ_1) can be definitely determined as the stress at this knee. For the above-mentioned three carbonate rocks, the yield stress could be determined as the stress required to produce a specified amount of permanent strain. For Mizuho trachyte,

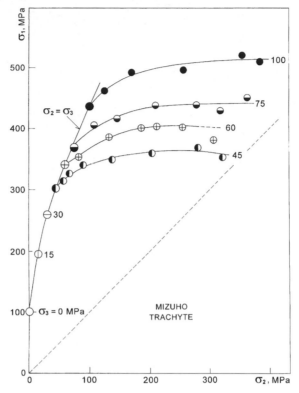

Figure 3.44.   Stress at fracture (σ_1) as functions of the intermediate principal stress (σ_2) in Mizuho trachyte. Different symbols show the different σ_3, indicated by numerals in MPa. Larger symbols show the case of $\sigma_2 = \sigma_3$.

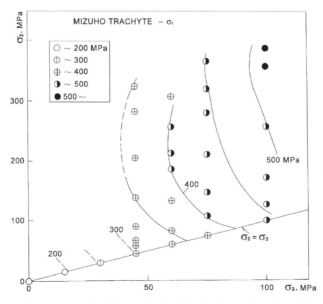

Figure 3.45. σ_1 at fracture as a function of σ_2 and σ_3 in the σ_2–σ_3 plane, in Mizuho trachyte.

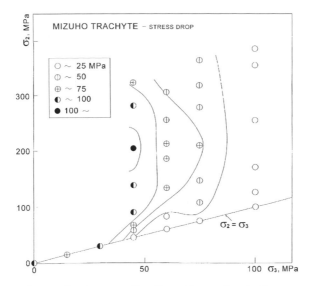

Figure 3.46.   Stress drop at fracture as a function of σ_2 and σ_3 in the σ_2–σ_3 plane, in Mizuho trachyte.

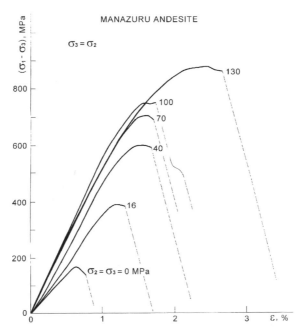

Figure 3.47.   Stress ($\sigma_1 - \sigma_3$) – strain (ε) curves by the conventional triaxial compression tests ($\sigma_1 > \sigma_2 = \sigma_3$) of Manazuru andesite. Numerals for each curve are confining pressure in MPa. ε: axial strain.

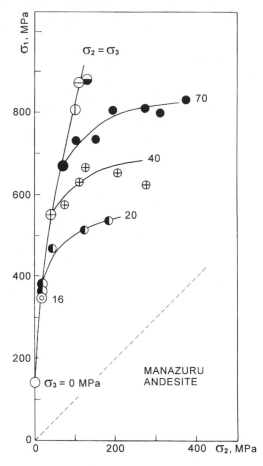

Figure 3.48.   Stress at fracture (σ_1) as functions of σ_2 in Manazuru andesite. Different symbols show the different σ_3, indicated by numerals in MPa. Larger symbols show the case of $\sigma_2 = \sigma_3$.

however, the bending of deformation curves is not sharp and so there is some ambiguity in the determination of yield stress. Therefore, the following discussion is limited to the three carbonate rocks. The marked changes in slope of each stress-strain curve were observed at a permanent strain of about 0.2–0.5% in these rocks. In Dunham dolomite, since a change in slope of each stress-strain curve is slightly gradual, the axial stress at 0.5% permanent strain was taken as the yield stress. In Solnhofen limestone and Yamaguchi marble, however, stresses at 0.2% offset were taken as the yield stress. This small difference in the permanent strain for determination of the yield points seems to have no significant effect.

3.3.f.1   *Dunham dolomite*

Figure 3.53 shows the maximum principal stress σ_1 at the yield point as a function of the intermediate principal stress σ_2 for various minimum principal stress σ_3 in Dunham dolomite. In this case, σ_1 increases clearly with the increase of σ_2.

Table 3.8. Summary of true triaxial compression tests of Manazuru andesite: Fracture.

Experiment No.	σ_1 (MPa)	σ_2 (MPa)	σ_3 (MPa)	$\dfrac{\sigma_1 + \sigma_3}{2}$ (MPa)	τ_{oct} (MPa)
224	140	0	0	70	66
213	349	16	16	183	157
228	364	20	20	192	157
226	381	20	20	201	170
225	470	67	20	245	202
223	516	124	20	268	214
227	538	186	20	279	216
212	552	40	40	296	241
215	577	75	40	309	245
220	632	112	40	336	264
211	669	126	40	355	278
209	653	206	40	347	259
216	626	278	40	333	241
202	671	70	70	371	283
206	735	101	70	403	306
201	735	152	70	403	296
208	808	193	70	439	323
207	812	275	70	441	313
204	801	313	70	436	304
205	833	375	70	452	314
218	806	100	100	453	333
221	875	110	110	493	361
219	881	130	130	506	354

Figure 3.54 shows the differential stress $(\sigma_1 - \sigma_3)$ at the yield point as a function of σ_2, for various σ_3 values. This figure shows that the $(\sigma_1 - \sigma_3)$ vs σ_2 curves for various σ_3 values agree very well with each other and can, in fact, be expressed by a single simple curve. This result shows that the yield strength $(\sigma_1 - \sigma_3)$ is independent of σ_3, but $(\sigma_1 - \sigma_3)$ increases markedly with the increase of σ_2. Subsequently, we can suggest the following empirical formula for yield stress:

$$(\sigma_1 - \sigma_3) = f(\sigma_2) \qquad \text{or} \qquad (3.3)$$

$$\sigma_1 = \sigma_3 + f(\sigma_2) \qquad (3.4)$$

where f is a monotonous function. The numerical data of this experiment are shown in Table 3.11.

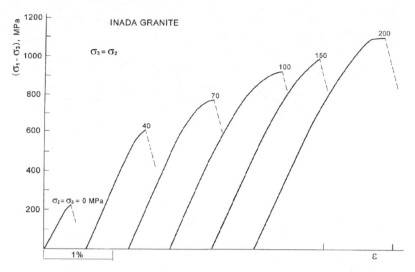

Figure 3.49.   Stress $(\sigma_1 - \sigma_3)$ – strain (ε) curves by the conventional triaxial compression tests $(\sigma_1 > \sigma_2 = \sigma_3)$ of Inada granite. Numerals for each curve are confining pressure in MPa. ε: axial strain.

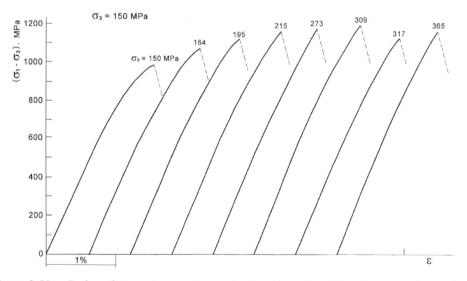

Figure 3.50.   Series of stress $(\sigma_1 - \sigma_3)$ – strain (ε) curves for different values of σ_2 under a constant σ_3 (=150 MPa) of Inada granite. Numerals for each curve are σ_2 values in MPa.

3.3.f.2 *Solnhofen limestone*

Figure 3.55 shows the maximum principal stress σ_1 at the yield point as a function of the intermediate principal stress σ_2, for various minimum principal stress σ_3. In the case of Solnhofen limestone, the effect of σ_2 is much smaller than in the case

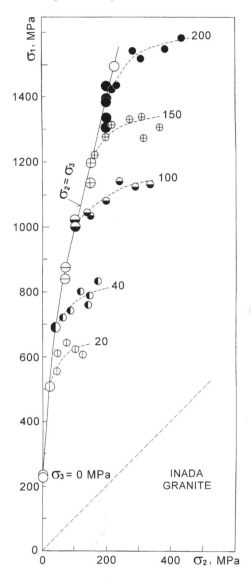

Figure 3.51. Stress at fracture (σ_1) as functions of σ_2 in Inada granite. Different symbols show the different σ_3, indicated by numerals in MPa. Larger symbols show the case of $\sigma_2 = \sigma_3$.

of Dunham dolomite (see the preceding section). Similarly, the effect of σ_3 on the fracture stress of Solnhofen limestone is much smaller than that observed in Dunham dolomite.

Figure 3.56 shows the differential stress ($\sigma_1 - \sigma_3$) at the yield point as a function of σ_2, for various σ_3 values. This figure shows that these ($\sigma_1 - \sigma_3$) vs σ_2 curves are nearly similar and that they can be expressed by a single curve. It is also clearly observable that the σ_2 effect is small in this rock. The numerical data of this experiment are shown in Table 3.12.

Table 3.9.   Summary of true triaxial compression tests of Inada granite: Fracture.

Experiment No.	σ_1 (MPa)	σ_2 (MPa)	σ_3 (MPa)	$\dfrac{\sigma_1 + \sigma_3}{2}$ (MPa)	τ_{oct} (MPa)
141	226	0	0	113	107
125	232	0	0	116	109
108	508	20	20	264	230
114	556	44	20	288	247
112	611	46	20	316	273
119	643	74	20	332	282
115	624	101	20	322	268
111	607	127	20	314	255
123	692	40	40	366	307
104	722	63	40	381	316
136	743	88	40	392	321
109	802	120	40	421	342
105	760	142	40	400	318
106	791	147	40	416	332
101	834	174	40	437	347
124	841	70	70	456	363
127	879	70	70	475	381
143	1003	100	100	552	426
148	1023	100	100	562	435
150	1046	138	100	573	437
155	1037	148	100	569	431
118	1083	199	100	592	442
157	1141	239	100	621	461
153	1125	291	100	613	445
149	1131	336	100	616	441
142	1138	150	150	644	466
144	1198	150	150	674	494
171	1223	164	150	687	503
172	1279	195	150	715	522
159	1316	215	150	733	535
140	1332	273	150	741	531
174	1342	309	150	746	528
173	1275	317	150	713	496
138	1307	365	150	729	502
120	1301	200	200	751	519
117	1398	200	200	799	565
128	1438	200	200	819	584
122	1388	200	200	794	560
129	1334	200	200	767	535
130	1425	216	200	813	574
145	1441	233	200	821	558
135	1545	281	200	873	616
147	1523	309	200	862	600
133	1554	383	200	877	600
134	1587	439	200	894	605
126	1497	230	230	864	597

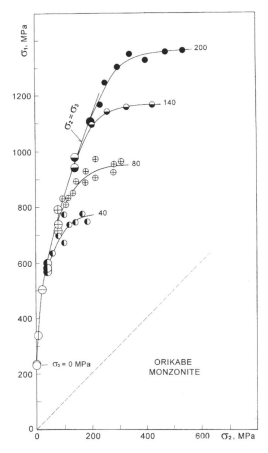

Figure 3.52.   Stress at fracture (σ_1) as functions of σ_2 in Orikabe monzonite. Different symbols show the different σ_3 values, indicated by numerals in MPa. Larger symbols show the case of $\sigma_2 = \sigma_3$.

3.3.f.3   *Yamaguchi marble*

Figure 3.57 shows the maximum principal stress σ_1 at the yield point as a function of the intermediate principal stress σ_2 for various minimum principal stress σ_3. In this case, σ_1 increases markedly with the increase of σ_2.

Figure 3.58 shows the differential stress ($\sigma_1 - \sigma_3$) at the yield point as a function of σ_2, for various σ_3 values. This figure shows that the ($\sigma_1 - \sigma_3$) vs σ_2 curves for various σ_3 values agree well with each other and can be expressed by a single curve. This result is very similar to that of Dunham dolomite, that is, the yield strength ($\sigma_1 - \sigma_3$) is independent of σ_3 and it increases markedly with the increase of σ_2. The numerical data of this experiment are shown in Table 3.13.

Figure 3.59 shows stress ($\sigma_1 - \sigma_3$)–strain (ε) curves for a constant $\sigma_2(=108\,\text{MPa})$ (a) and for a constant $\sigma_3(=55\,\text{MPa})$ (b). When σ_2 is constant (left), although the fracture strength ($\sigma_1 - \sigma_3$) varies clearly with different σ_3, the shape of the stress-strain

Table 3.10. Summary of true triaxial compression tests of Orikabe monzonite: Fracture

Experiment No.	σ_1 (MPa)	σ_2 (MPa)	σ_3 (MPa)	$\dfrac{\sigma_1 + \sigma_3}{2}$ (MPa)	τ_{oct} (MPa)
304	236	0	0	118	111
306	232	0	0	116	109
307	339	5	5	172	157
305	504	20	20	262	228
303	583	40	40	312	256
254	571	40	40	306	250
252	600	40	40	320	264
262	636	59	40	338	277
255	698	80	40	369	301
260	673	101	40	357	285
258	775	102	40	408	333
261	739	121	40	390	312
251	747	143	40	394	323
256	777	168	40	409	322
257	748	187	40	394	305
302	718	80	80	399	301
202	742	80	80	411	312
211	794	80	80	437	337
205	834	95	80	457	352
208	810	108	80	445	338
212	836	117	80	458	348
308	854	135	80	467	353
203	893	147	80	487	368
207	889	182	80	485	360
204	930	183	80	505	379
309	906	216	80	493	362
210	973	218	80	527	393
209	926	281	80	503	361
310	956	284	80	518	374
206	966	311	80	523	375
301	943	140	140	542	379
454	981	140	140	561	396
451	1098	205	140	619	437
456	1144	259	140	642	448
455	1161	331	140	651	443
453	1168	424	140	654	433
407	1107	200	200	654	428
409	1168	235	200	684	448
413	1244	251	200	722	481
406	1305	298	200	753	499
415	1352	343	200	776	513
411	1329	401	200	765	492
405	1358	473	200	779	494
401	1364	537	200	782	489

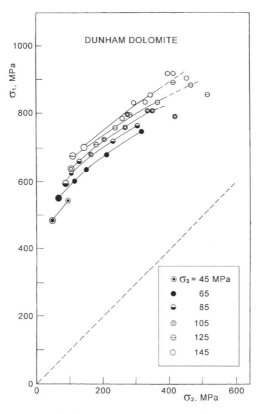

Figure 3.53. The maximum principal stress (σ_1) at the yield point as functions of σ_2 in Dunham dolomite. Different symbols show the different minimum principal stress σ_3. Larger symbols show the case of $\sigma_2 = \sigma_3$.

curves does not change for various σ_3 values, that is, the yield strength ($\sigma_1 - \sigma_3$) is independent of σ_3. When σ_3 is constant (right), the shape of the stress-strain curves changes markedly and the yield stress increases appreciably with increase of σ_2. This figure directly shows the effects of σ_2 and σ_3, which are quite different. Thus, it was clearly shown that the intermediate principal stress σ_2 plays an important role in rock deformation. For the yield stress of all three the above-mentioned isotropic homogeneous carbonate rocks, Equation (3.3) or (3.4) can be applied.

3.3.g *Failure criteria of rocks*

3.3.g.1 *Previous studies*
In general, the stresses at fracture and yielding can be expressed by

$$\sigma_1 = f(\sigma_2, \sigma_3) \qquad \text{or}$$
$$f'(\sigma_1, \sigma_2, \sigma_3) = 0 \tag{3.5}$$

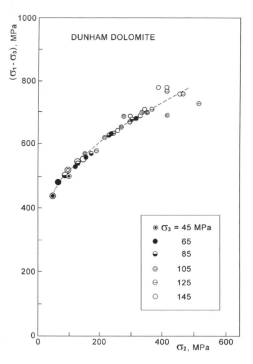

Figure 3.54.   Differential stress $(\sigma_1 - \sigma_3)$ at the yield point as a function of σ_2 in Dunham dolomite for various σ_3 values. Relation between the yield strength $(\sigma_1 - \sigma_3)$ and σ_2 is independent of σ_3, but $(\sigma_1 - \sigma_3)$ increases markedly with the increase of σ_2.

As mentioned above, the stress states, which result in the fracture or yielding of rocks, have been investigated by many researchers. In particular, the following fracture criteria have been widely discussed:

$$\text{Coulomb fracture criterion} \qquad \tau = \tau_0 + \mu_i\ \sigma_n \qquad\qquad (3.6)$$

$$\text{or} \quad \sigma_1 = \sigma_c + \mu_i'\ \sigma_3$$

$$\text{Mohr fracture criterion} \qquad \tau = f''(\sigma_n) \qquad\qquad (3.7)$$

$$\text{or} \quad \sigma_1 = \sigma_c + f'''(\sigma_3)$$

$$\text{Griffith criterion} \qquad (\sigma_1 - \sigma_3)^2 = 8\sigma_T(\sigma_1 + \sigma_3)$$
$$(\text{if} \quad \sigma_1 - 3\sigma_3 > 0) \qquad\qquad (3.8)$$

Griffith criterion modified by McClintock and Walsh (1962)

$$(\sigma_1 - \sigma_3)(1 + \mu)^{1/2} + \mu(\sigma_1 + \sigma_3) = 4\sigma_T \qquad\qquad (3.9)$$

Table 3.11. Summary of true triaxial compression tests on Dunham dolomite (block 2): Yield.

Experiment No.	σ_1 (MPa)	σ_2 (MPa)	σ_3 (MPa)	$\dfrac{\sigma_1 + \sigma_2 + \sigma_3}{3}$ (MPa)	τ_{oct} (MPa)
103	485	45	45	192	207
107	545	97	45	229	224
113	535	60	60	218	224
124	550	65	65	227	229
140	600	117	65	261	241
114	635	153	65	284	251
112	680	208	65	318	262
115	745	318	65	376	280
145	593	85	85	254	239
153	627	128	85	280	245
136	660	153	85	299	256
143	720	233	85	346	271
135	765	306	85	385	283
131	635	105	105	281	250
120	680	167	105	317	257
121	725	205	105	345	271
134	760	268	105	378	278
141	800	270	105	392	296
108	810	334	105	416	293
142	810	356	105	423	292
138	795	415	105	438	282
172	675	125	125	308	259
158	705	187	125	325	260
174	760	239	125	374	276
119	800	293	125	406	286
176	835	362	125	441	295
106	895	414	125	478	317
171	885	463	125	491	310
105	855	516	125	499	298
159	700	145	145	330	262
130	790	253	145	396	282
173	835	296	145	425	296
148	835	324	145	435	292
146	855	343	145	448	299
133	925	392	145	487	325
157	925	410	145	493	324
123	905	455	145	502	312
152	805	200	200	402	285

where τ is the shear stress, σ_n is the normal stress, σ_c is the uniaxial compressive strength, σ_T is the tensile strength, μ, μ_i and μ_i' are the constants related to sliding friction, and f–f‴ are monotonically increasing functions. As mentioned before, the experimental results obtained from the conventional triaxial compression test

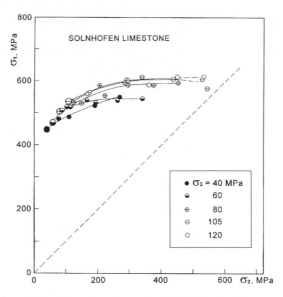

Figure 3.55. σ_1 at the yield point as functions of σ_2 in Solnhofen limestone. Different symbols show the different σ_3 values. Larger symbols show the case of $\sigma_2 = \sigma_3$.

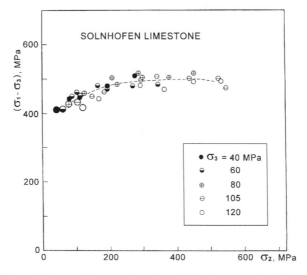

Figure 3.56.   Differential stress $(\sigma_1 - \sigma_3)$ at the yield point as a function of σ_2 in Solnhofen limestone for various σ_3 values.

$(\sigma > \sigma_2 = \sigma_3)$ are well explained by the Coulomb-Mohr criterion (and also the modified Griffith theory). However, although the σ_2 effect is also important, these current criteria do not take this effect into account.

Among failure criteria that consider the σ_2 effect, the high applicability of the von Mises yield criterion for ductile metals has been established by extensive

Table 3.12. Summary of true triaxial compression tests on Solnhofen limestone: Yield

Experiment No.	σ_1 (MPa)	σ_2 (MPa)	σ_3 (MPa)	$\dfrac{\sigma_1 + \sigma_2 + \sigma_3}{3}$ (MPa)	τ_{oct} (MPa)
210	448	40	40	176	192
203	446	40	40	175	191
206	481	80	40	200	199
205	485	113	40	213	195
201	520	193	40	251	200
204	547	274	40	287	207
249	469	60	60	196	193
256	508	87	60	218	205
247	519	102	60	227	207
254	514	113	60	229	202
243	540	164	60	254	206
251	535	197	60	264	200
246	538	275	60	291	195
252	543	345	60	316	199
232	507	80	80	222	201
240	535	126	80	247	204
244	529	150	80	253	197
236	583	208	80	290	216
233	552	225	80	285	197
234	595	283	80	319	211
238	584	298	80	320	206
231	583	378	80	347	207
222	594	454	80	376	216
230	536	105	105	249	203
235	565	185	105	285	200
225	600	293	105	333	204
226	612	341	105	353	207
229	604	439	105	383	207
227	604	528	105	412	220
228	577	547	105	410	215
211	534	120	120	258	195
216	560	168	120	282	197
215	600	294	120	338	198
212	586	362	120	356	190
213	611	450	120	394	204
214	614	533	120	422	216

studies (Nadai, 1950). This criterion is expressed by use of the octahedral shear stress (τ_{oct}) as follows:

$$\tau_{oct} = \text{constant}$$

$$\tau_{oct} \equiv 1/3\{(\sigma_1 - \sigma_2)^2 + (\sigma_2 - \sigma_3)^2 + (\sigma_3 - \sigma_1)^2\}^{1/2} \qquad (3.10)$$

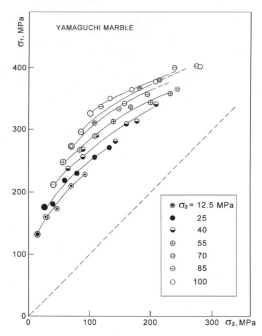

Figure 3.57. σ_1 at the yield point as functions of σ_2 in Yamaguchi marble. Different symbols show the different σ_3 values. Larger symbols show the case of $\sigma_2 = \sigma_3$.

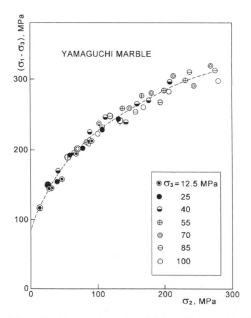

Figure 3.58.   Differential stress $(\sigma_1 - \sigma_3)$ at the yield point as a function of σ_2 in Yamaguchi marble for various σ_3 values. Yield strength $(\sigma_1 - \sigma_3)$ is independent of σ_3, but $(\sigma_1 - \sigma_3)$ increases markedly with the increase of σ_2.

Table 3.13. Summary of true triaxial compression tests on Yamaguchi marble: Yield

Experiment No.	σ_1 (MPa)	σ_2 (MPa)	σ_3 (MPa)	$\dfrac{\sigma_1 + \sigma_2 + \sigma_3}{3}$ (MPa)	τ_{oct} (MPa)
152	81	0	0	27	38
153	113	6	6	42	50
145	130	12.5	12.5	52	55
148	158	26	12.5	66	65
144	158	28	12.5	66	65
150	172	45	12.5	77	69
142	209	67	12.5	96	83
146	227	90	12.5	110	88
105	175	25	25	75	71
104	180	39	25	81	70
107	218	58	25	100	83
101	228	78	25	110	86
102	255	107	25	129	95
103	270	132	25	142	100
110	210	40	40	97	80
111	236	64	40	113	87
112	255	88	40	128	92
114	267	88	40	132	98
113	288	112	40	147	104
115	280	143	40	154	98
118	307	160	40	169	109
116	312	177	40	176	111
117	339	208	40	196	122
119	246	55	55	119	90
120	266	82	55	134	94
121	288	107	55	150	100
122	311	138	55	168	108
123	334	167	55	185	115
124	342	200	55	199	117
125	356	231	55	214	124
154	272	70	70	137	95
141	309	108	70	162	105
136	331	148	70	183	109
139	353	181	70	201	116
143	378	214	70	221	126
140	363	243	70	225	120
138	393	268	70	244	133
155	295	85	85	155	99
137	335	119	85	180	111
132	340	157	85	194	107
135	354	194	85	211	111
133	398	238	85	240	128

(*Continued*)

Table 3.13.   (*Continued*)

Experiment No.	σ_1 (MPa)	σ_2 (MPa)	σ_3 (MPa)	$\dfrac{\sigma_1 + \sigma_2 + \sigma_3}{3}$ (MPa)	τ_{oct} (MPa)
134	401	274	85	253	130
129	324	100	100	175	106
128	344	133	100	192	108
130	362	169	100	210	111
126	385	207	100	231	118
127	400	279	100	260	123
159	398	150	150	233	117
160	397	150	150	232	116
164	453	200	200	284	119
165	454	200	200	285	120

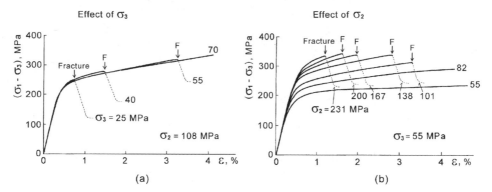

Figure 3.59.   (a) shows the effects of σ_3 under the constant σ_2 (=108 MPa) and (b) shows the effect of σ_2 under the constant σ_3 (=55 MPa) in Yamaguchi marble.

This criterion can be interpreted as specifying that yielding will occur when the distortional strain energy reaches a constant value. The von Mises theory was also interpreted by Nadai (1950) as the criterion that fracture will occur when the octahedral shear stress (τ_{oct}) reaches a critical value.

Lode (1928) and Nadai (1950) also proposed the following formula.

$$\tau_{oct} = f(\sigma_{oct}) \tag{3.11}$$

where
$$\sigma_{oct} \equiv 1/3(\sigma_1 + \sigma_2 + \sigma_3).$$

This criterion, that the octahedral shear stress (τ_{oct}) at fracture is a monotonically increasing function of the mean stress (σ_{oct}), was proposed in order to explain the pressure-dependency of the strength of brittle materials. Various formulae of this type were proposed by several researchers. For example,

$$\tau_{oct} = C_1\sigma_{oct} + C_2 \tag{3.12}$$

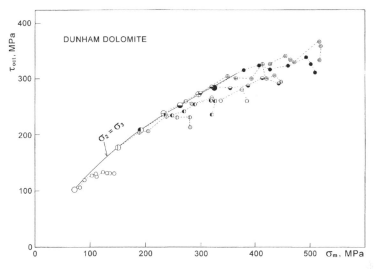

Figure 3.60. Examination of fracture criterion proposed by Lode (1928) and Nadai (1950). Experimental data of fracture stress of Dunham dolomite are widely scattered in the τ_{oct} vs $\sigma_{oct}(=\sigma_m)$ graph.

was applied by Bresler and Pister (1957) for concrete. The relation

$$\tau_{oct}^2 = C\sigma_{oct} \tag{3.13}$$

was discussed by Murrell (1965) as the formal extension of the Griffith theory to three dimensions. Sakurai and Serata (1967) proposed the following equation,

$$\tau_{oct} = f(\alpha\sigma_{oct} + \beta\sigma_3 + \gamma) \tag{3.14}$$

where α, β and γ are constants.

However, the examination by the experimental data obtained for rocks mentioned before shows that these fracture criteria fail to correlate with the experimental results. Figure 3.60 shows a case of examination of fracture criterion proposed by Lode (1928), Nadai (1950) and others. Experimental data of fracture stress of Dunham dolomite were plotted in the τ_{oct} vs σ_{oct} graph (Mogi, 1970b). Circles in this figure are widely scattered. This result indicates that this fracture criterion fails to correlate with the experimental data obtained from the true triaxial compression test. Other formulae mentioned above also fail to correlate with the experimental data.

Wiebols and Cook (1968) theoretically proposed a hypothesis that the fracture strength is determined by effective shear strain energy while considering the strain energy around Griffith cracks by the frictional sliding of crack surfaces. Figure 3.61 shows the effect of σ_2 and σ_3 on the fracture strength calculated by this effective shear strain energy theory. These curves depend on the assumed values of the coefficient of sliding friction. The curves are for the frictional coefficient equal to 1.0. It is noteworthy that the pattern of these theoretical curves seems to be fairly similar to the abovementioned results obtained experimentally using the true triaxial compression tests.

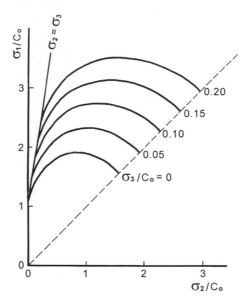

Figure 3.61.   The effect of σ_2 and σ_3 on the fracture strength σ_1/C_0 according to the effective shear strain energy criterion (Wiebols and Cook, 1968). Internal frictional coefficient $= 1.0$ is assumed and C_0 is uniaxial compressive strength.

As for the yield stress of rocks, perhaps, no yield criterion has been presented to date. This might be caused by that reliable experimental data on rocks and concrete were extremely scarce.

3.3.g.2   *Fracture criterion*

As mentioned above, the von Mises criterion had been extensively confirmed by a number of experiments on yield stress of ductile metals. It has been suggested that a satisfactory criterion of fracture of brittle materials might be obtained by a modification of the von Mises criterion. According to this criterion, yielding occurs when the distortional strain energy (τ_{oct}) reaches a constant value. For brittle fracture, however, the distortional strain energy at fracture is not constant, because brittle fracture is characterized by high pressure dependency. Accordingly, as mentioned in the preceding section, $\tau_{oct} = f(\sigma_{oct})$ was proposed by several investigators as fracture criterion. However, as shown in Fig. 3.60, criteria of this type fail to correlate with the experimental date.

As will be clearly shown in the later section, the shear fault surface at fracture is a nearly flat plane parallel to the σ_2 direction. Therefore, it was supposed that the critical distortional strain energy increased with effective mean normal stress $1/2 \, (\sigma_1 + \sigma_3)$. Thus, the following criterion was proposed by the present author.

$$\tau_{oct} = f_1(\sigma_1 + \sigma_3) \tag{3.15}$$

where f_1 is a monotonically increasing function.

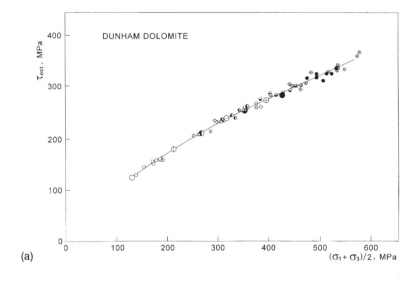

(a)

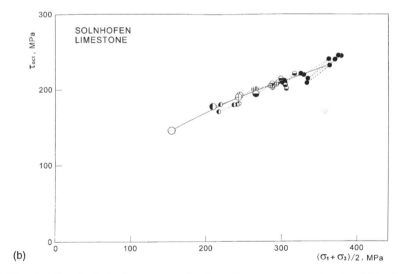

(b)

Figure 3.62. (a) Octahedral shear stress (τ_{oct}) at fracture versus $(\sigma_1 + \sigma_3)/2$ of Dunham dolomite. Different symbols show different σ_3. (the same in (b)–(g)); (b) Solnhofen limestone; (c) Yamaguchi marble; (d) Mizuho trachyte; (e) Manazuru andesite; (f) Inada granite; (g) Orikabe monzonite.

In Figs. 3.62(a)–(g), τ_{oct} at fracture is plotted against $(\sigma_1 + \sigma_3)/2$, based on the experimental results for Dunham dolomite, Solnhofen limestone, Yamaguchi marble, Mizuho trachyte, Manazuru andesite, Inada granite, and Orikabe monzonite (Table 3.4–3.10). Different symbols correspond to different σ_3 values and larger symbols show the case of $\sigma_2 = \sigma_3$. As the circles for the true triaxial stress state form a single

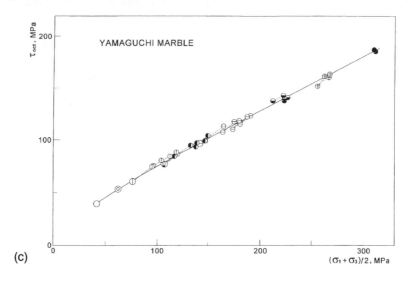

(c)

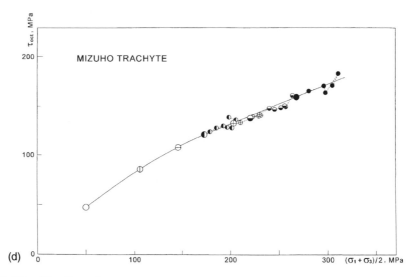

(d)

Figure 3.62.   (*Continued*)

curve for each rock, the high applicability of the new fracture criterion (Equation 3.15) seems to be well established.

In Fig. 3.63 τ_{oct}/C_0 is plotted against $(\sigma_1 + \sigma_3)/2C_0$ in logarithmic scale, where C_0 is the uniaxial compressive strength. Since these curves for various rocks are nearly linear, the empirical formula may be expressed approximately by the following power function:

$$\tau_{oct} \approx A(\sigma_1 + \sigma_3)^n \qquad (3.16)$$

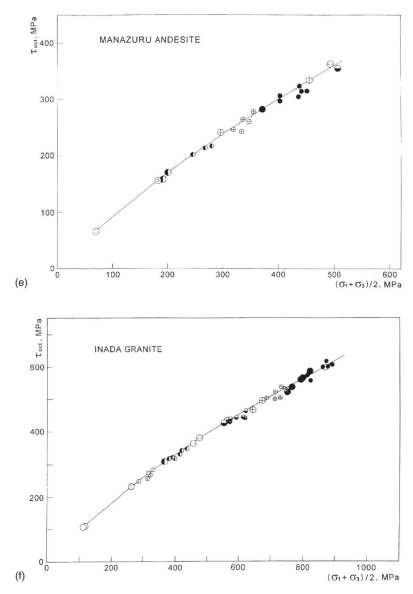

Figure 3.62.   (*Continued*)

where A and n are certain empirical constants. Values of the constant n for several rocks were obtained from this figure as follows (Mogi, 1972):

Solnhofen limestone	0.56
Dunham dolomite	0.72
Yamaguchi marble	0.74
Inada granite	0.87
Manazuru andesite	0.88

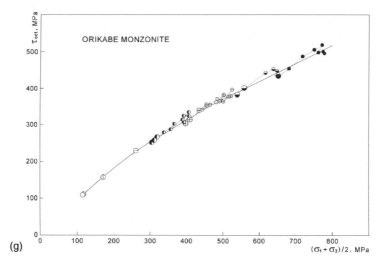

(g)

Figure 3.62.   (*Continued*)

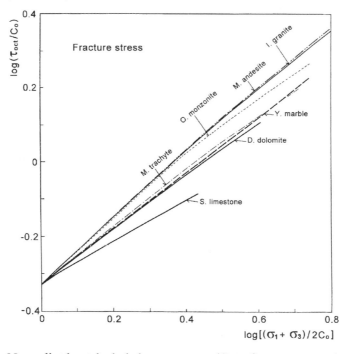

Figure 3.63.   Normalized octahedral shear stress τ_{oct}/C_0 at fracture versus $(\sigma_1 + \sigma_3)/2C_0$ in logarithmic scale, of 3 carbonate rocks and 4 silicate rocks.

Recently, Chang and Haimson (2000b) also showed that their data obtained by the true triaxial compression tests supported Equation 3.16, and they obtained n values as follows:

Westerly granite   0.89
KTB amphibolite   0.86

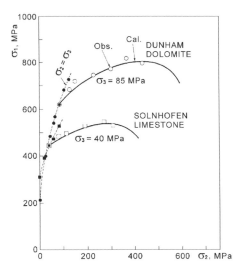

Figure 3.64. σ_1 at fracture as a function of σ_2 for different σ_3 values in Dunham dolomite (circles) and Solnhofen limestone (squares). Solid lines are curves calculated from the fracture criterion $\tau_{oct} = f(\sigma_1 + \sigma_3)$, on the basis of the conventional triaxial compression data (solid symbols and broken lines). (Mogi, 1971c)

On the basis of the above mentioned experiments in 1970's, it was concluded that states of stress, which produce fracture, might be expressed by Equation 3.15 and 3.16 obtained by generalization of the von Mises criterion for yielding of ductile metals. The physical interpretation of the new fracture criterion may be stated as follows: fracture will occur when the distortional strain energy reaches a critical value which increases with the effective mean normal stress on the fault plane. Rock fracture under the true triaxial compression occurs by a shear faulting, of which the direction is always parallel to the σ_2 direction, and so the mean normal pressure on the fault plane is $(\sigma_1 + \sigma_3)/2$.

However, the fault plane is not exactly flat and parallel to the σ_2 direction. Therefore, the formula (Equation 3.15) may be modified into the following form:

$$\tau_{oct} = f(\sigma_1 + \sigma_3 + \alpha\sigma_2) \tag{3.17}$$

where α is a small constant. The difference between the formula (3.15) and (3.17) was difficult to be found from the actual data because of their dispersion.

To examine the applicability of the empirical fracture criterion (Equation 3.15), the curves calculated from $\tau_{oct} = f(\sigma_1 + \sigma_3)$ were compared with the experimental data obtained using the true triaxial test (Fig. 3.64). In this figure, circular and square symbols correspond to experimental data of Dunham dolomite and Solnhofen limestone, respectively. Solid and open symbols correspond to the data obtained using the conventional triaxial test and the true triaxial test, respectively. Using the broken curves obtained using the conventional triaxial compression test ($\sigma_2 = \sigma_3$), the solid

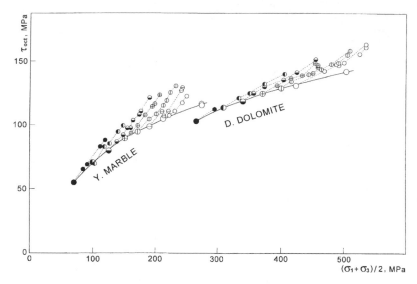

Figure 3.65.   Octahedral shear stress τ_{oct}/C_0 at yield point versus $(\sigma_1 + \sigma_3)/2$ for Yamaguchi marble and Dunham dolomite. Different symbols correspond to different confining pressure (σ_3). Circles for different stress states are widely scattered.

curves for the true triaxial compression are calculated on the basis of the fracture criterion $\tau_{oct} = f(\sigma_1 + \sigma_3)$. The observed data and the calculated curves agree well in the investigated range of σ_2 values. Thus, Equation 3.15 is applicable as the empirical fracture criterion (Mogi, 1971c). However, the applicability of Equation 3.15 may decrease for high σ_2 values. This is perhaps due to the fact that the fault plane is not exactly flat and parallel to the σ_2 direction. In such case, Equation 3.17 may be useful. This is a problem which should be investigated further.

3.3.g.3 *Yield criterion*

As mentioned before, the high applicability of the von Mises yield criterion for ductile metals has been established by extensive studies. The von Mises criterion is applied to cases where the yield stress is independent of the confining pressure. However, the yield stress of rocks increases with increasing confining pressure in many cases (e.g., Paterson, 1967). Therefore, a pressure-dependent criterion is needed for yielding of rocks, similarly as it was for brittle fracturing. In this case, also, it is probable that a satisfactory criterion may be obtained by the generalization of the von Mises criterion. From this viewpoint, various criteria were examined based on the results of the true triaxial compression tests.

In Fig. 3.65, τ_{oct} at yielding of Dunham dolomite and Yamaguchi marble is plotted against $(\sigma_1 + \sigma_3)/2$, as was done for fracture. Different symbols correspond to differ-ent confining pressure (σ_3) and larger symbols show the case where $\sigma_2 = \sigma_3$. Circles for different stress states are widely scattered for both rocks. This result shows that the fracture criterion (Equation 3.15) cannot be applied to yielding.

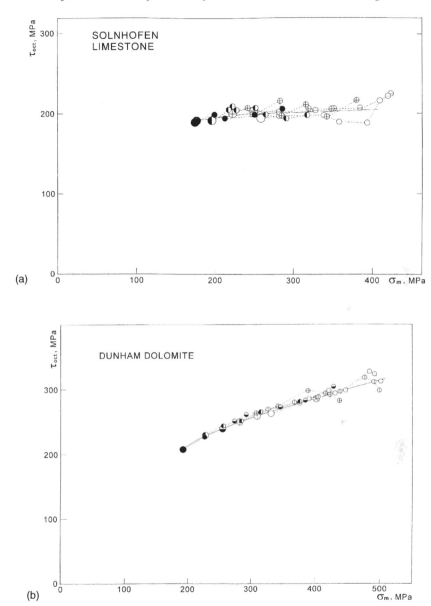

Figure 3.66.   (a) Octahedral shear stress (τ_{oct}) at yield point is plotted against the mean stress $(\sigma_1 + \sigma_2 + \sigma_3)/3$ for Solnhofen limestone. Different symbols corresponding to different σ_3 form nearly a single line; (b) Dunham dolomite; (c) Yamaguchi marble.

In Figs. 3.66 (a)–(c), τ_{oct} at yielding is plotted against the mean stress $(\sigma_1 + \sigma_2 + \sigma_3)/3$ for Dunham dolomite, Solnhofen limestone and Yamaguchi marble. Data points obtained from the true triaxial compression tests form a single curve for each rock. Thus, $\tau_{oct} = f(\sigma_1 + \sigma_2 + \sigma_3)$ (Equation 3.11), which was proposed by Nadai (1950) and others as fracture criteria, is satisfactory as the yield criterion for rocks.

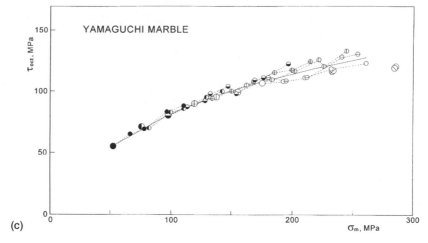

(c)

Figure 3.66.   (*Continued*)

In these figures, the slope of the $\tau_{oct} \sim (\sigma_1 + \sigma_2 + \sigma_3)$ curve is different for different rocks, and it shows the degree of the pressure-dependency of yielding. In Solnhofen limestone, the slope of the $\tau_{oct} \sim (\sigma_1 + \sigma_2 + \sigma_3)$ curve is very small and the curve is nearly parallel to the $(\sigma_1 + \sigma_2 + \sigma_3)$ axis, namely, τ_{oct} is constant for various stress states, and so the von Mises criterion can be applied approximately to this rock, as it is to most ductile metals.

3.3.g.4   *Summary*

Current failure criteria of rocks are divided into the following two groups: The first does not take into account the effect of the intermediate principal stress σ_2 (for example, the Coulomb-Mohr criterion) and the second considers this effect. According to the true triaxial compression test results mentioned above, the σ_2 effect is important in many cases and so no satisfactory criteria can be obtained without considering the σ_2 effect.

Based on the true triaxial compression test of the seven rocks mentioned above, the following failure criteria were proposed, by generalization of the von Mises yield criterion:

$$\text{Fracture}: \quad \tau_{oct} = f_1(\sigma_1 + \sigma_3) \tag{3.18}$$

$$\text{Yield}: \quad \tau_{oct} = f_2(\sigma_1 + \sigma_2 + \sigma_3) \tag{3.19}$$

where τ_{oct} is the octahedral shear stress defined as

$$\tau_{oct} \equiv 1/3[(\sigma_1 - \sigma_2)^2 + (\sigma_2 - \sigma_3)^2 + (\sigma_3 - \sigma_1)^2]^{1/2}, \tag{3.20}$$

and f_1 and f_2 are monotonically increasing functions. That is, τ_{oct} is not always constant, but increases monotonically with $(\sigma_1 + \sigma_3)$ for fracture and $(\sigma_1 + \sigma_2 + \sigma_3)$ for yielding.

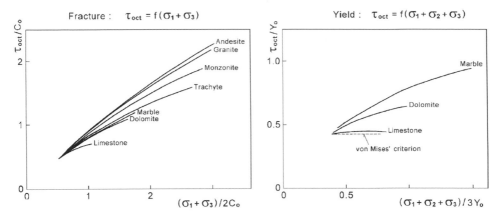

Figure 3.67. Summary of stress states at fracture and yielding of tested rocks (expressed by the generalized von Mises criterion). C_0 is the uniaxial compressive strength, Y_0 is the yield stress at the ductility of 1%.

The physical interpretation of the criteria may be stated as follows: Fracture or yielding of rocks will occur when the distortional strain energy reaches a critical value. This critical energy is not constant, but monotonically increases with effective mean normal stress. In fracture, a shear faulting takes place in a plane parallel to the σ_2 direction, and so the effective normal stress may be independent of σ_2. On the other hand, yielding does not occur on such macroscopic slip planes in a definite direction, so the mean stress $(\sigma_1 + \sigma_2 + \sigma_3)/3$ is taken as the effective mean normal stress. Thus, the difference between fracture and yielding may be reasonably interpreted.

The failure conditions of tested rocks are summarized in Fig. 3.67, where C_0 is uniaxial compressive strength and Y_0 is yield stress at the ductility of 1%, obtained using the conventional test (Mogi, 1971b). $\tau_{oct} \sim (\sigma_1 + \sigma_3)$ and $\tau_{oct} \sim (\sigma_1 + \sigma_2 + \sigma_3)$ curves are nearly linear but concave slightly downward. The slopes of these curves are different for different rocks and they show the degree of the pressure-dependency of fracture strength and yield strength. The von Mises yield criterion is included in the generalized yield criterion as a special case with zero slope.

3.3.h *Ductility, fracture pattern and dilatancy*

3.3.h.1 *Ductility and stress drop*
Figure 3.68 shows stress-strain curves of Dunham dolomite and the pressure dependence of ductility of rock in the conventional triaxial test ($\sigma_2 = \sigma_3$). Ductility (ε_n), measured by permanent strain before fracture, increases nearly linearly with increasing confining pressure. Figure 3.69 shows stress-strain curves at various values of the intermediate principal stress (σ_2) and the minimum principal stress (σ_3) of 125 MPa for Dunham dolomite (left figure) and the ductility for various values of σ_3 as functions of σ_2 (right figure). The ductility increases markedly with increasing σ_3, but decreases appreciably with increasing σ_2, as mentioned before.

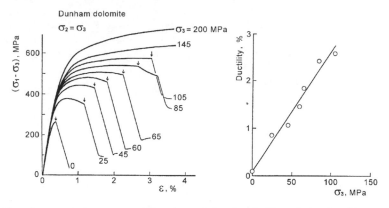

Figure 3.68.   Stress-strain curves of Dunham dolomite (left) and the ductility - σ_3 relation (right) in the conventional triaxial test ($\sigma_2 = \sigma_3$).

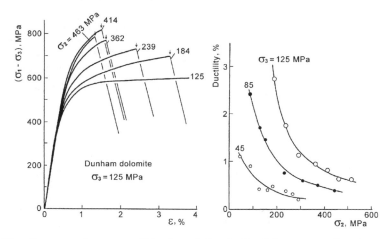

Figure 3.69.   Stress-strain curves of Dunham dolomite for different σ_2 values ($\sigma_3 = 125$ MPa) (left) and the ductility – σ_2 relation for various σ_3 values in the truc triaxial test (right).

In Fig. 3.70, the ductility (ε_n) under the true triaxial compression is plotted against ($\sigma_3 - \alpha\sigma_2$) for Yamaguchi marble, Dunham dolomite and Solnhofen limestone, where different symbols correspond to different σ_3 values in each plot. This figure suggests that the relation between ε_n and ($\sigma_3 - \alpha\sigma_2$) can be expressed approximately by a single monotonic curve for different σ_2 and σ_3 values.

In Fig. 3.71, the logarithm of ductility (log ε_n) of Yamaguchi marble and Dunham dolomite is plotted against the quantity ($\sigma_3 - \alpha\sigma_2$). Although circles somewhat scatter, they can be fitted fairly well with a single line. Thus, the following empirical formula may be applicable:

$$\log \varepsilon_n = K_1(\sigma_3 - \alpha\sigma_2) + K_2 \qquad (3.21)$$

where α, K_1 and K_2 are constants. The value of α is 0.25 for Yamaguchi marble, and 0.3 for Dunham dolomite and 0.05 for Solnhofen limestone. Thus, it is concluded

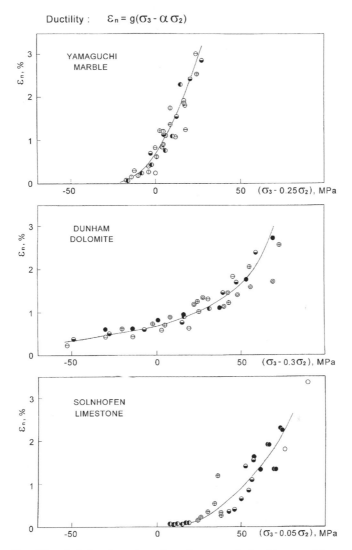

Figure 3.70. Ductility (ε_n) is plotted against ($\sigma_3 - \alpha\sigma_2$) for Yamaguchi marble, Dunham dolomite and Solnhofen limestone. Different symbols correspond to different σ_3 values.

that the effects of σ_2 and σ_3 on ductility are opposite, and that the magnitude of the σ_2 effect on ductility is significantly different for different rocks.

The coefficient of strain-hardening (h), defined as the slope of the linear part of the stress-strain curve in the post-yield region, shows a noticeable feature. In Fig. 3.72, the coefficient of strain-hardening h of Yamaguchi marble and Solnhofen limestone is plotted against σ_2. Different symbols show different σ_3. All circles can be nearly fitted with a single curve. That is, the coefficient of strain-hardening increases monotonically with increasing σ_2, but is nearly independent of σ_3. It is a very striking result that the shape of stress-strain curve before fracture, determined mainly by the

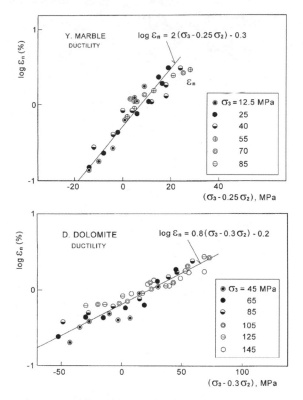

Figure 3.71. $\log \varepsilon_n$ of Yamaguchi marble and Dunham dolomite is plotted against $(\sigma_3 - \alpha\sigma_2)$, and fitted with a linear function.

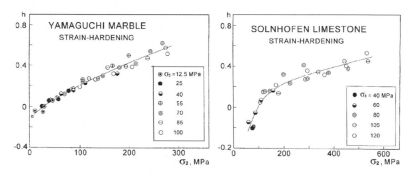

Figure 3.72.   Coefficient of strain-hardening (h), defined as the slope of the stress-strain curve in the post-yield region, of Yamaguchi marble and Solnhofen limestone is plotted against σ_2. This result shows that h monotonically increases with increasing σ_2 and is independent σ_3 (Mogi, 1972a).

coefficient of strain-hardening, is markedly affected by the intermediate principal stress σ_2 and independent of the minimum principal stress σ_3 (Mogi, 1972a).

Figure 3.73 shows the residual stress (S_R) at fracture of Yamaguchi marble for different σ_3 values as functions of σ_2. S_R increases markedly with increasing σ_3,

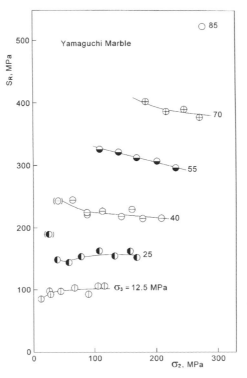

Figure 3.73. Residual stress ($S_R = \mu\sigma_n$) at fracture of Yamaguchi marble for different σ_3 values as a function of σ_2. μ: sliding friction coefficient.

but does not change significantly with increasing σ_2. From this result and the angles between the σ_1-axis and the fracture plane which is parallel to the σ_2-direction, the sliding friction coefficient (μ) on the fracture plane can be obtained. In Fig. 3.74, μ values for different σ_3 values are plotted against σ_2. Except for a very low σ_3 value (=12.5 MPa), μ values are nearly constant (=0.92) in this rock.

In Fig. 3.75 (a) (b), fracture stress (σ_1), residual stress (S_R), friction coefficient (μ), stress drop (S_D) and ductility (ε_n) of Yamaguchi marble and Dunham dolomite are shown as functions of the intermediate principal stress (σ_2) under constant σ_3 conditions. The mechanical behavior shown in this figure is nearly similar in these two rocks.

3.3.h.2 *Fracture pattern*

One of the most important results of the true triaxial compression tests which were carried out by the present author in 1970's is that the brittle fracture of isotropic and homogeneous rocks occur by shear faulting and the fracture surface is a nearly flat plane which is parallel to the σ_2 direction. An example of the tested rock specimen with steel end pieces and silicon rubber jacket is presented in Fig. 3.76.

The fracture patterns of the tested specimens are shown in detail in several photographs. Figure 3.77 (top figure) shows schematically the tested specimen, in which

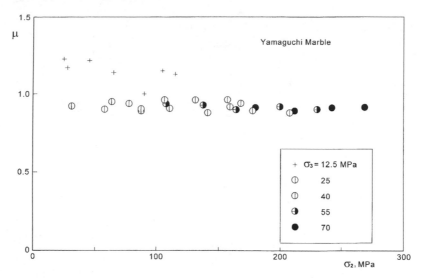

Figure 3.74.   Sliding friction coefficient (μ) of Yamaguchi marble is nearly constant except for a very low σ_3.

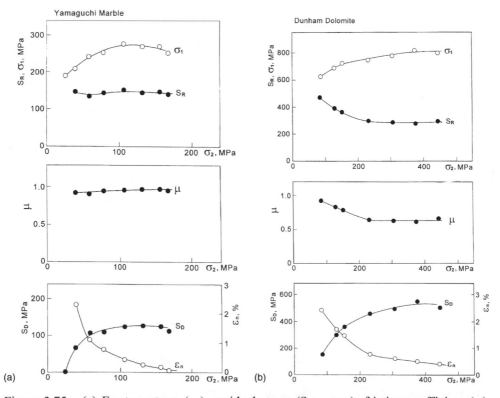

Figure 3.75.   (a) Fracture stress (σ_1), residual stress ($S_R = \mu\sigma_n$), friction coefficient (μ), stress drop ($S_D = \sigma_1 - \mu\sigma_n$) and ductility ($\varepsilon_n$) for Yamaguchi marble as functions of σ_2, for $\sigma_3 = 25$ MPa; (b) The same as (a) for Dunham dolomite for $\sigma_3 = 85$ MPa.

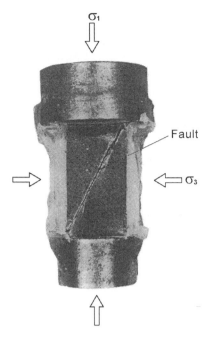

Figure 3.76.　A rock tested using the true triaxial compression apparatus (Mogi, 1969).

shear faulting occurred. The fault plane is parallel to the σ_2 direction and the angle between the fault plane and the direction of the maximum principal stress (σ_1) is indicated by the *fracture angle (θ)* which is dependent on the σ_2 values. In the following photographs, the two planes A and B perpendicular to the σ_2 direction on both sides of rectangular parallelepiped rock specimen are shown, and in some cases the fracture patterns in the central part of both sides (C and D) are shown at an enlarged scale.

　　Figure 3.77 (bottom figure) shows two examples of a schematic sketch of typical fracture patterns of Dunham dolomite. The main faults are quite linear, and many fine cracks are parallel to the main fault and another fine cracks are conjugate to the main fault. Corresponding to the unevenness of the main fault, microcrack density changes locally. This pattern suggests the local stress distribution along the fault. As it can be seen in the left figure, the main fault sometimes does not pass near the corner of the rock sample, and both the directions of the main fault and many lineations of microcracks agree very well. This result shows that there is no significant end effect, and therefore the fracture angle (θ) determined by the direction of the main fault is reliable.

3.3.h.2.1　Dunham dolomite

Dunham dolomite is an ideally homogeneous-isotropic and fine-grained carbonate rock. Fortunately, since the color of this rock is deep gray or black, the main fault, as well as fine microcracks, can be clearly observed.

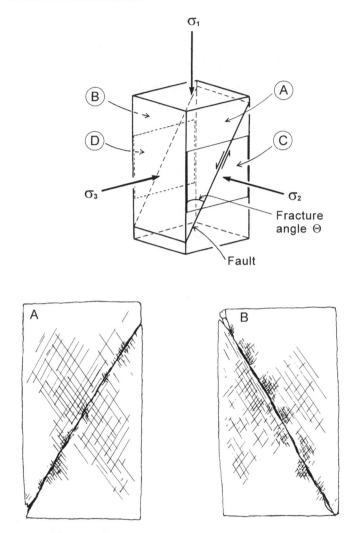

Figure 3.77.   Top and bottom figures show schematically the tested rock specimen in which shear faulting occurred. Bottom figures show two examples of a schematic sketch of typical fracture patterns of Dunham dolomite.

Figure 3.78 (a) shows the case of sample No. 123 which is a typical one. The main fault traces in A and B planes are inclined straight lines which are completely symmetric. This result shows that the fault is a nearly flat plane parallel to the σ_2 direction. It is noted that the conjugate linear patterns of microcracks are also clearly observed and one of the conjugate crack lineations is parallel to the main fault trace. From the directions of the main fault trace and the microcrack lineation, the fracture angle between the fault plane and σ_1 axis can be determined accurately. Such simple fracture patterns are seen in Figs. 3.78 (d), 3.78 (g), 3.78 (i), 3.78 (k) and 3.78 (l).

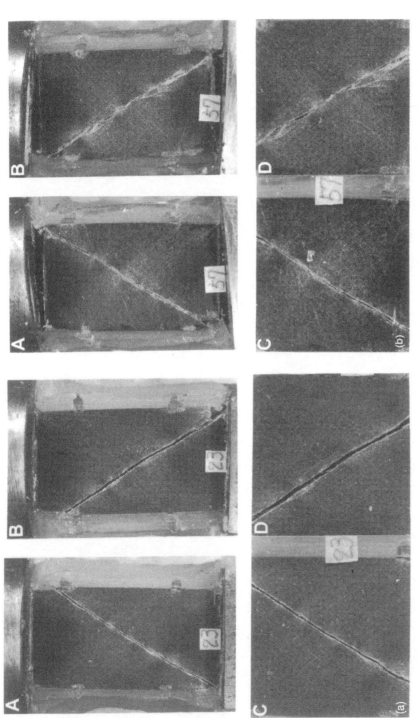

Figure 3.78. Fracture patterns of Dunham dolomite (a)–(l). (a) Dunham dolomite No. 123. $\sigma_1 = 952$ MPa, $\sigma_2 = 455$ MPa, $\sigma_3 = 145$ MPa; (b) Dunham dolomite No. 157. $\sigma_1 = 1002$ MPa, $\sigma_2 = 410$ MPa, $\sigma_3 = 145$ MPa; (c) Dunham dolomite No. 133. $\sigma_1 = 1015$ MPa, $\sigma_2 = 392$ MPa, $\sigma_3 = 145$ MPa; (d) Dunham dolomite No. 146. $\sigma_1 = 922$ MPa, $\sigma_2 = 349$ MPa, $\sigma_3 = 145$ MPa; (e) Dunham dolomite No. 148. $\sigma_1 = 923$ MPa, $\sigma_2 = 324$ MPa, $\sigma_3 = 145$ MPa; (f) Dunham dolomite No. 173. $\sigma_1 = 927$ MPa, $\sigma_2 = 296$ MPa, $\sigma_3 = 145$ MPa; (g) Dunham dolomite No. 130. $\sigma_1 = 883$ MPa, $\sigma_2 = 253$ MPa, $\sigma_3 = 145$ MPa; (h) Dunham dolomite No. 106. $\sigma_1 = 941$ MPa, $\sigma_2 = 414$ MPa, $\sigma_3 = 125$ MPa; (i) Dunham dolomite No. 119. $\sigma_1 = 863$ MPa, $\sigma_2 = 293$ MPa, $\sigma_3 = 125$ MPa; (j) Dunham dolomite No. 121. $\sigma_1 = 786$ MPa, $\sigma_2 = 205$ MPa, $\sigma_3 = 105$ MPa; (k) Dunham dolomite No. 171. $\sigma_1 = 918$ MPa, $\sigma_2 = 463$ MPa, $\sigma_3 = 125$ MPa; (l) Dunham dolomite No. 108. $\sigma_1 = 824$ MPa, $\sigma_2 = 334$ MPa, $\sigma_3 = 105$ MPa.

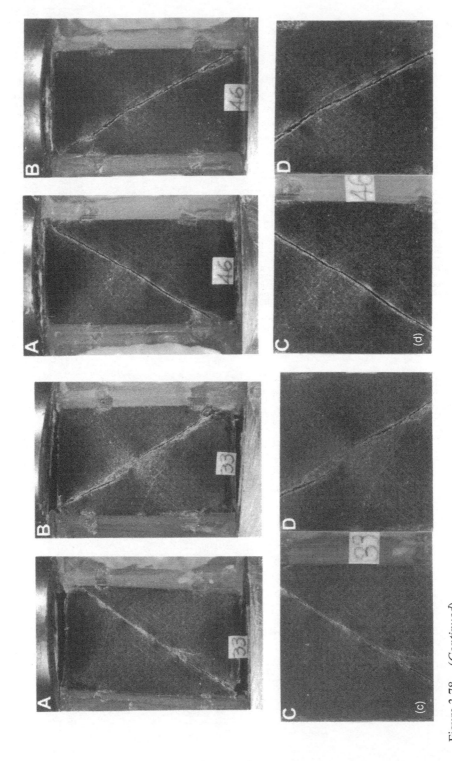

Figure 3.78.   *(Continued)*

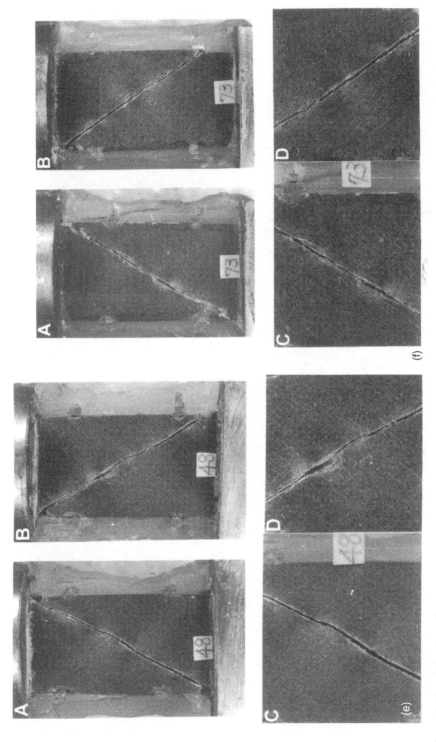

Figure 3.78.   *(Continued)*

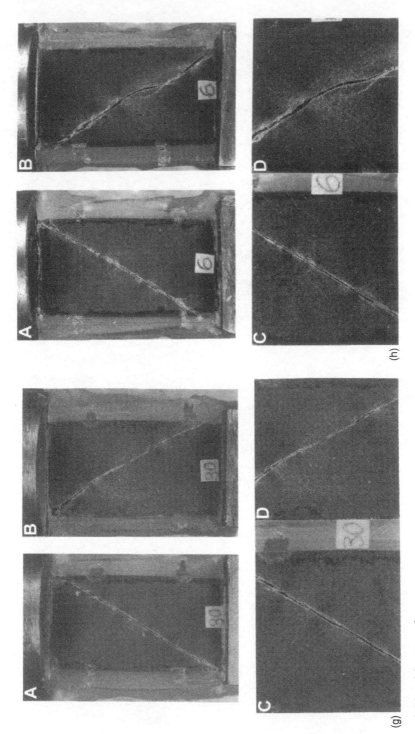

Figure 3.78.   (*Continued*)

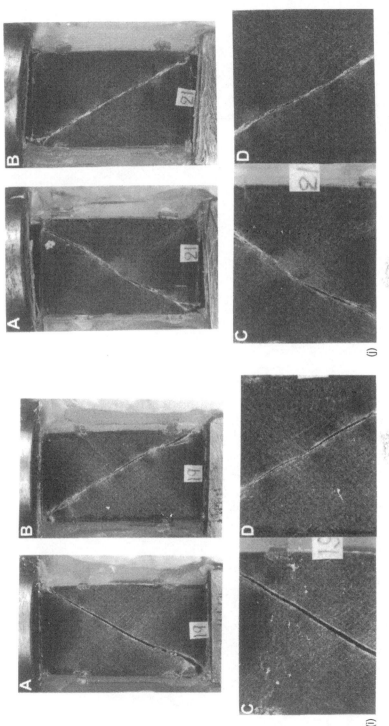

Figure 3.78.   (*Continued*)

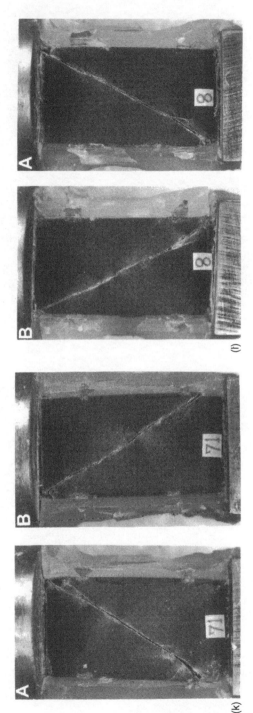

Figure 3.78.   *(Continued)*

DUNHAM DOLOMITE

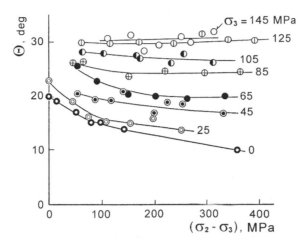

Figure 3.79.   Fracture angle (θ) of Dunham dolomite under true triaxial compression as functions of the intermediate principal stress ($\sigma_2 - \sigma_3$) for different σ_3 values.

However, there are some cases in which the fracture patterns are more complex. In the case of Fig. 3.78 (b), the trace of the main fault in plane A is simple, but it is slightly shifted in the central region. On the other hand, the main fault in plane B is complex and includes several branches of moderate cracks. In the case of Fig. 3.78 (c), the main fault trace in plane A is a straight line, but the main fault trace in plane B is shifted significantly in the central region. These complex fracture patterns may be attributed to the effect of the underdeveloped conjugate fault. A similar shift of the main fault can also be seen in Figs. 3.78 (j) and 3.78 (l). The distribution of microcracks is particularly interesting, because the high density region of microcracks may probably correspond to the highly stressed region.

Although the fault traces are not always completely linear as mentioned above, from these many experimental results, it is concluded that the fault plane of Dunham dolomite under true triaxial compression is fundamentally a nearly flat plane parallel to the σ_2 direction.

Figure 3.79 shows the fracture angle (θ) of Dunham dolomite under true triaxial compression as a function of the intermediate principal stress σ_2. Different symbols correspond to different σ_3 values. The fracture angle increases with increasing σ_3, but decreases with increasing σ_2, particularly under lower σ_3 values.

3.3.h.2.2   Solnhofen limestone

Solnhofen limestone is also an ideally homogeneous-isotropic and very-fine-grained carbonate rock. This is an ivory-white colored rock. Typical fracture patterns under true triaxial compression are shown in Figs. 3.80(a)–(e). The main fault traces in planes A and B are inclined straight lines and completely symmetric. In this rock, however,

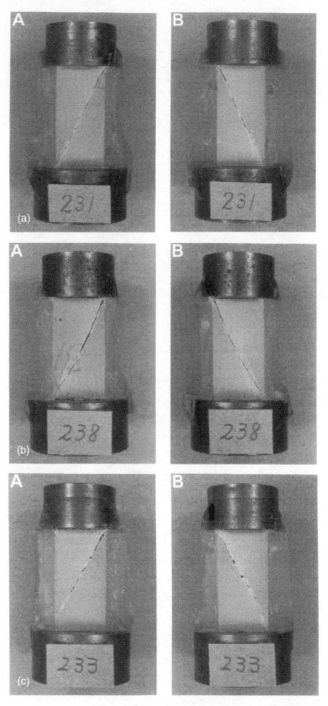

Figure 3.80.   Fracture patterns of Solnhofen limestone (a)–(e). (a) Solnhofen limestone No. 231 $\sigma_1 = 650$ MPa, $\sigma_2 = 378$ MPa, $\sigma_3 = 80$ MPa; (b) Solnhofen limestone No. 238 $\sigma_1 = 665$ MPa, $\sigma_2 = 298$ MPa, $\sigma_3 = 80$ MPa; (c) Solnhofen limestone No. 233 $\sigma_1 = 591$ MPa, $\sigma_2 = 225$ MPa, $\sigma_3 = 80$ MPa; (d) Solnhofen limestone No. 246 $\sigma_1 = 553$ MPa, $\sigma_2 = 275$ MPa, $\sigma_3 = 60$ MPa; (e) Solnhofen limestone No. 244 $\sigma_1 = 577$ MPa, $\sigma_2 = 150$ MPa, $\sigma_3 = 80$ MPa.

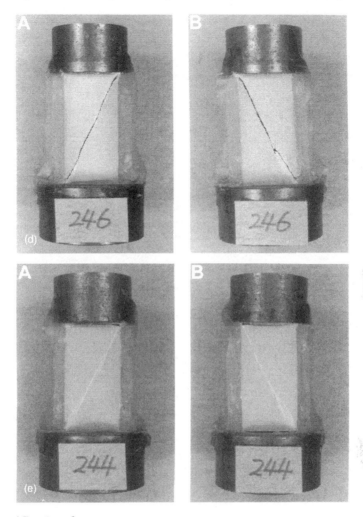

Figure 3.80. (*Continued*)

no significant microcrack pattern, which was observed in Dunham dolomite, can be recognized. This may be not attributed to the color of this rock, but to the mechanical property due to its very-fine-grained structure. It should be noted, however, that if the fault traces are examined more carefully, the fault trace is not a simple straight line, but an *en echelon* pattern is more or less recognizable. This result suggests that the fault plane is not a completely flat plane, but has various kinds of asperities of small scale.

In any case, these figures strongly support the conclusion that the fault plane is a nearly flat plane parallel to the σ_2 direction, as is seen in Dunham dolomite. The fracture angle (θ) can be determined accurately in many cases. Figure 3.81 shows θ as function of σ_2, where different symbols correspond to different σ_3 values. The

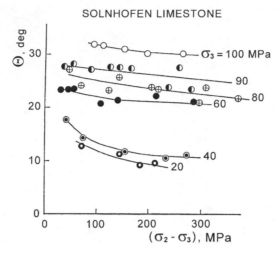

Figure 3.81.   Fracture angle (θ) of Solnhofen limestone under true triaxial compression as functions of the intermediate principal stress ($\sigma_2 - \sigma_3$) for different σ_3 values.

fracture angle (θ) increases with increasing σ_3, but decreases with increasing σ_2, particularly under lower σ_3 values. The σ_2-dependency of the fracture angle of Solnhofen limestone is nearly similar to that of Dunham dolomite.

3.3.h.2.3   Yamaguchi marble
Yamaguchi marble is a homogeneous, isotropic and fine-grained white marble. Figure 3.82 (a) shows two examples of ruptured specimens under true triaxial compression conditions. The typical shear fault trace can be seen clearly. In this rock, a significant microcrack pattern is not observed. Generally, the fault trace of the white marble is not clear. To emboss the fracture pattern, the rock specimens, on which thin, brittle, black-colored films were pasted on planes A and B, were used for the true triaxial experiments. Figs. 3.82(b)–(d) show several examples of the fracture patterns. In these cases, also, the main traces in planes A and B are inclined and symmetric. Although the fault traces sometimes slightly curve or branch, this result shows that the fault plane of Yamaguchi marble under true triaxial compression is fundamentally a nearly flat plane parallel to the σ_2 direction.

Figure 3.83 shows the fracture angle (θ) of Yamaguchi marble as a function of σ_2 and σ_3. Fracture angle slightly increases with increasing σ_3, but slightly decreases, or does not change, with increasing σ_2. In this rock, the fracture angles are relatively large (25°–32°) and so changes in fracture angle are small.

3.3.h.2.4   Mizuho trachyte
Mizuho trachyte is a *macroscopically* homogeneous, isotropic and medium-grained volcanic rock. Figures 3.84 (a)–(d) show several examples of specimens ruptured under true triaxial compression conditions. In these cases, the main fault traces in planes A and B are inclined, roughly straight lines (or a band) and symmetric. These

Figure 3.82. Fracture patterns of Yamaguchi marble (a_1 and a_2–d). (a_1) No. 102 $\sigma_1 = 275$ MPa, $\sigma_2 = 107$ MPa, $\sigma_3 = 25$ MPa; (a_2) No. 115 $\sigma_1 = 307$ MPa, $\sigma_2 = 143$ MPa, $\sigma_3 = 40$ MPa; (b) Yamaguchi marble No. 134 $\sigma_1 = 537$ MPa, $\sigma_2 = 274$ MPa, $\sigma_3 = 85$ MPa; (c) Yamaguchi marble No. 124 $\sigma_1 = 396$ MPa, $\sigma_2 = 200$ MPa, $\sigma_3 = 55$ MPa; (d) Yamaguchi marble No. 144 $\sigma_1 = 177$ MPa, $\sigma_2 = 28$ MPa, $\sigma_3 = 12.5$ MPa.

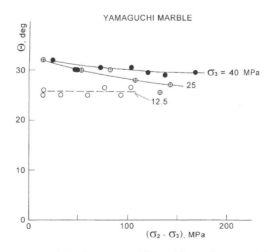

Figure 3.83. Fracture angle (θ) of Yamaguchi marble under true triaxial compression as functions of the intermediate principal stress ($\sigma_2 - \sigma_3$) for different σ_3 values.

Figure 3.84.   Fracture patterns of Mizuho trachyte (a)–(d). (a) Mizuho trachyte No. 323 $\sigma_1 = 440\,\text{MPa}$, $\sigma_2 = 279\,\text{MPa}$, $\sigma_3 = 75\,\text{MPa}$; (b) Mizuho trachyte No. 322 $\sigma_1 = 381\,\text{MPa}$, $\sigma_2 = 306\,\text{MPa}$, $\sigma_3 = 60\,\text{MPa}$; (c) Mizuho trachyte No. 315 $\sigma_1 = 403\,\text{MPa}$, $\sigma_2 = 212\,\text{MPa}$, $\sigma_3 = 60\,\text{MPa}$; (d) Mizuho trachyte No. 318 $\sigma_1 = 401\,\text{MPa}$, $\sigma_2 = 186\,\text{MPa}$, $\sigma_3 = 60\,\text{MPa}$.

results show that the fault is a roughly flat plane parallel to the σ_2 direction. In some cases, a noticeable conjugate fracture pattern (e.g., Fig. 3.84 (d)) can be seen. Thus, the fault plane may be not a simple flat plane, and it has many small asperities which may be related to grain boundaries and other mechanical heterogeneities. However, in this rock also, it may be concluded that the fault plane is fundamentally a roughly flat plane parallel to the σ_2 direction.

3.3.h.2.5   Inada granite

Inada granite is a *macroscopically* homogeneous, isotropic and coarse-grained igneous hard rock. Figures 3.85(a)–(d) show several examples of specimens ruptured under true triaxial compression conditions. In some cases, the fracture pattern is complex and the main fault is not completely linear. However, from these experimental results, it may be concluded that the main fault traces in planes A and B are inclined, roughly straight lines and symmetric, so the fault plane of Inada granite under true triaxial compression is fundamentally a roughly flat plane parallel to the σ_2 direction.

3.3.h.2.6   Orikabe monzonite

Orikabe monzonite is also a *macroscopically* homogeneous, isotropic and coarse-grained igneous hard rock. Figures 3.86 (a) (b) shows two examples of specimens ruptured under true triaxial compression conditions. The fracture patterns of this rock include the main simple fault trace and complex minor cracks. The fault trace of sample No. 410 is quite linear, but that of No. 414 is slightly curved. The main feature of the fracture pattern of this rock is essentially similar to those of Inada granite and other rocks, mentioned above.

3.3.h.3   *Dilatancy*

The inelastic deformation which is a combination of plastic deformation and microfracturing increases before fracture in many rocks. Plastic deformation does not produce volumetric strain, but deformation related to microfracturing causes an increase in volumetric strain because microfracturing in brittle rocks includes crack opening (e.g., Matsushima, 1960; Brace, Paulding and Scholz, 1966). Based on these experimental results, the dilatancy models of earthquake generation processes have been proposed by a number of researchers (e.g., Nur, 1972; Scholz, Sykes and Aggarwal, 1973; Mjachkin, Brace, Sobolev and Dietrich, 1975; Mogi, 1974). Until that time, the dilatancy had been measured under the conventional triaxial stress state in which two of the principal stresses are equal.

Mogi (1977) measured the dilatancy under the true triaxial stresses states. The methods of strain measurement in the directions of the three principal stresses are shown in Fig. 3.87 (b) (c). ε_1, ε_2 and ε_3 in this section are defined as the strains in the directions of σ_1, σ_2 and σ_3 axes, respectively, due to the increase of the axial stress $(\sigma_1 - \sigma_3)$ under the constant σ_2 and σ_3 $[(\varepsilon_1 = \varepsilon_2 = \varepsilon_3 = 0$ for $(\sigma_1 - \sigma_3) = 0)]$. The volumetric strain due to the increase of the axial differential stress $(\sigma_1 - \sigma_3)$ was

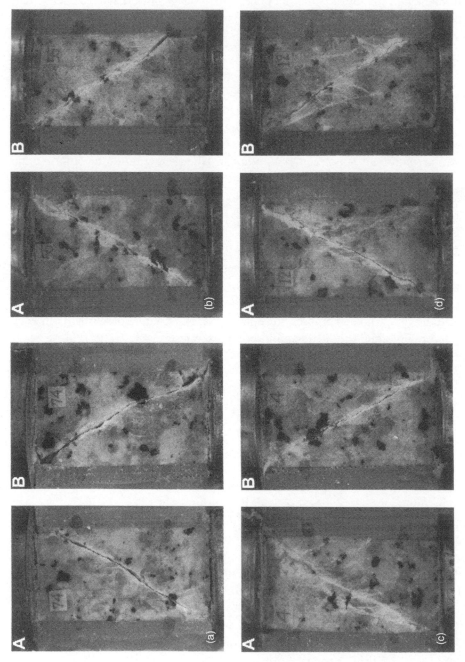

Figure 3.85.   Fracture patterns of Inada granite (a)–(d). (a) Inada granite No. 174 $\sigma_1 = 1342$ MPa, $\sigma_2 = 309$ MPa, $\sigma_3 = 150$ MPa; (b) Inada granite No. 155 $\sigma_1 = 1037$ MPa, $\sigma_2 = 148$ MPa, $\sigma_3 = 100$ MPa; (c) Inada granite No. 104 $\sigma_1 = 722$ MPa, $\sigma_2 = 63$ MPa, $\sigma_3 = 40$ MPa; (d) Inada granite No. 112 $\sigma_1 = 611$ MPa, $\sigma_2 = 46$ MPa, $\sigma_3 = 20$ MPa.

Figure 3.86. Fracture patterns of Orikabe monzonite. (a) Orikabe monzonite No. 410 $\sigma_1 = 1396$ MPa, $\sigma_2 = 387$ MPa, $\sigma_3 = 230$ MPa; (b) Orikabe monzonite No. 414 $\sigma_1 = 1353$ MPa, $\sigma_2 = 499$ MPa, $\sigma_3 = 230$ MPa.

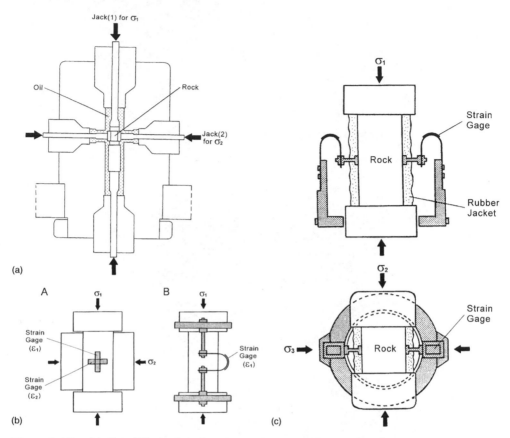

Figure 3.87.   (a) Simplified schematic figure of the true triaxial compression cell and a rock sample inside the cell; (b) Methods of the measurements of axial deformation (ε_1) and the lateral deformation (ε_2). (see text 2.1.a and Fig. 2.2.); (c) Method of the measurement of the lateral deformation in the direction of minimum compression (σ_3-axis). Upper: front section; lower: horizontal section.

calculated using the following equation:

$$\Delta V/V_0 = \varepsilon_1 + \varepsilon_2 + \varepsilon_3 \qquad (3.22)$$

where ΔV and V_0 are the volume change and the original volume of the specimen, respectively. It should be noted that this equation is applicable only to the case where deformation occurs nearly uniformly within the specimen.

3.3.h.3.1   Mizuho trachyte
In Fig. 3.88, the differential stress ($\sigma_1 - \sigma_3$) – strain (ε_1) curves under $\sigma_3 = 35\,\text{MPa}$ are shown for different σ_2 values. In Fig. 3.89, the differential stress ($\sigma_1 - \sigma_3$) (top graphs) and the lateral strains ε_2 and ε_3 (bottom graphs) are shown as functions of

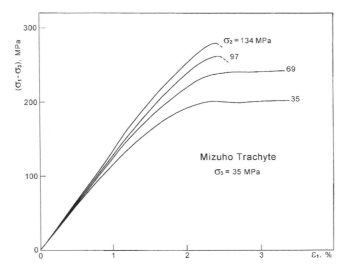

Figure 3.88. Differential stress $(\sigma_1 - \sigma_3)$ – strain (ε_1) curves under $\sigma_3 = 35$ MPa for different σ_2 in Mizuho trachyte.

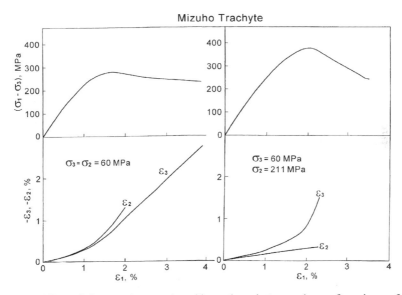

Figure 3.89. Differential stress $(\sigma_1 - \sigma_3)$ and lateral strains ε_2 and ε_3 as functions of the axial strain ε_1 in Mizuho trachyte. Left: axi-symmetric stress state $(\sigma_2 = \sigma_3)$; right: the case of true triaxial compression.

the axial strain ε_1. The left figure shows the case of the axi-symmetric stress state $(\sigma_1 = \sigma_3)$. In this case, two lateral strains $(\varepsilon_2$ and $\varepsilon_3)$ are nearly equal. The right figure shows a typical case of the true triaxial compression test $(\sigma_2 = 211$ MPa and $\sigma_3 = 60$ MPa). In this case, the ultimate strength $(\sigma_1 - \sigma_3)_{max}$ is higher and the stress

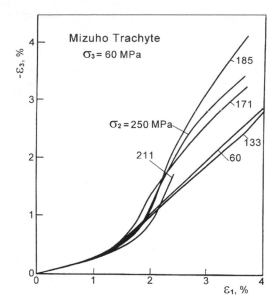

Figure 3.90.   Lateral strains ε_3 in the direction of the minimum compression σ_3 as functions of the axial strain ε_1 in Mizuho trachyte. Numerals for each curve are σ_2 values in MPa.

decrease after yielding is larger than those in the case of the conventional triaxial compression (shown in the left figure), and it is noted that ε_3 is much greater than ε_2.

Figure 3.90 shows ε_3 as functions of ε_1 under the true triaxial stress state. In this case, σ_3 is constant (60 MPa), and numerals for each curve are σ_2 in MPa. Plots are linear in the initial stage and concave upward in the ε_1 range from 1–2%. In a later stage, ε_3 seems to increase markedly with increase of σ_2.

Figure 3.91 shows ε_2 as functions of ε_1. Numerals for each curve are σ_2 in MPa. The curves are concave upward for lower σ_2 values, but nearly linear for higher σ_2 values. ε_2 decreases markedly with increase of σ_2.

In Fig. 3.92, the volumetric strain $\Delta V/V_0$ is shown as functions of ε_1 which increases monotonously with σ_1 until a stage of rupture. Numerals for each curve are σ_2 in MPa, and σ_3 is held at 60 MPa. The volumetric strain decreases linearly in the initial stage, and the curves are concave upward at some value of ε_1. As discussed by Matsushima (1960) and Brace, Paulding and Scholz (1966), it is deduced that the deformation in the linear part of these curves is elastic and the deviation from the elastic curve in a later stage is due to the dilatancy (the increase in the inelastic volumetric strain). In this figure, the dilatancy begins to occur at a higher axial strain ε_1 for higher σ_2 values. Since ε_1 increases monotonously with σ_1 until the stage of rupture, as mentioned above, this result shows that the axial stress level (σ_1) at onset of the dilatancy increases with the intermediate principal stress σ_2.

Figure 3.93 shows the ratio of ε_3 to ε_2 as functions of the axial strain (ε_1). This ratio increases markedly with increasing σ_2, particularly in the inelastic region.

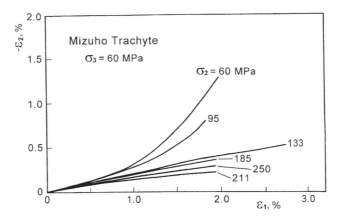

Figure 3.91.   Lateral strains ε_2 in the direction of the intermediate principal stress σ_2 as functions of the axial strain ε_1 in Mizuho trachyte. Numerals for each curve are σ_2 values in MPa.

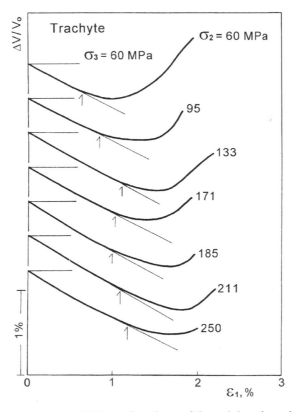

Figure 3.92.   Volumetric strain $\Delta V/V_0$ as functions of the axial strain ε_1 in Mizuho trachyte. Numerals for each curve are σ_2 values in MPa.

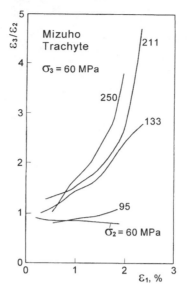

Figure 3.93.   Ratio of ε_3 and ε_2 as functions of the axial strain ε_1 in Mizuho trachyte. Numerals for each curve are σ_2 values in MPa.

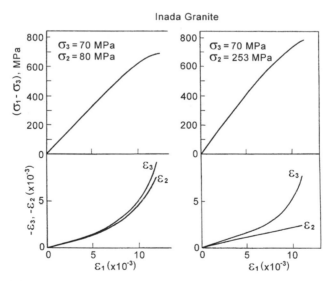

Figure 3.94.   Differential stress $(\sigma_1 - \sigma_3)$ and lateral strains ε_2 and ε_3 as functions of the axial strain ε_1 in Inada granite. Left: nearly axi-symmetric stress state; right: the case of true triaxial compression.

3.3.h.3.2   Inada granite

Figure 3.94 shows typical deformation curves of Inada granite which is a typical crystalline hard rock. In this case, the minimum principal stress σ_3 is 70 MPa. The curves in the upper figures show the differential stress $(\sigma_1 - \sigma_3)$ as functions of the axial strain (ε_1). In the lower figures, the lateral strains ε_2 and ε_3 are shown as

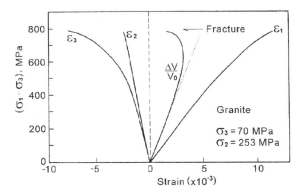

Figure 3.95. Three principal strains (ε_1, ε_2 and ε_3) and volumetric strain $\Delta V/V_0$ as functions of differential stress ($\sigma_1 - \sigma_3$) in Inada granite.

functions of ε_1. The left figure shows the case of the nearly axi-symmetric stress state ($\sigma_3 = 70$ MPa and $\sigma_2 = 80$ MPa). In this case, strains ε_2 and ε_3 are nearly equal and increase markedly before fracture. The right figure shows the case of true triaxial compression in which σ_2 is 253 MPa and σ_3 is 70 MPa. In this case, ε_3 increases markedly before fracture, but ε_2 curve is nearly linear until fracture.

Figure 3.95 shows the strains as functions of the differential stress ($\sigma_1 - \sigma_3$) in the case where $\sigma_2 = 253$ MPa and $\sigma_3 = 70$ MPa. The vertical axis is stress ($\sigma_1 - \sigma_3$) and the horizontal axis is strain. The three principal strains (ε_1, ε_2 and ε_3) and the volumetric strain ($\Delta V/V_0$) are shown by solid curves. The dotted line corresponds to the elastic volumetric strain (compaction). The dilatancy curve (the curve of inelastic volumetric strain) under true triaxial compressive stress state is similar to the case of the conventional triaxial stress state, but the result shows that the dilatancy is mainly caused by the increase of the strain ε_3 (in the direction of the minimum principal stress σ_3).

3.3.h.3.3 Yamaguchi marble

Figure 3.96 presents the case of Yamaguchi marble which shows a marked ductility under high pressure. In this case, the minimum principal stress σ_3 is 25 MPa. The left figure shows the case of the conventional triaxial stress state ($\sigma_2 = \sigma_3$). In this case, the two principal lateral strains are also equal ($\varepsilon_2 = \varepsilon_3$). The right figure shows a typical case of the true triaxial stress state ($\sigma_2 = 105$ MPa and $\sigma_3 = 25$ MPa). In this case, ε_3 is much greater than ε_2, particularly in the inelastic region.

Figure 3.97 shows the lateral strain ε_2 as functions of the axial strain ε_1. These curves are nonlinear for lower σ_2 values, but linear for higher σ_2 values, as seen in Mizuho trachyte and Inada granite. ε_2 decreases markedly with increase of σ_2. The change in the volumetric strain of this rock is nearly the same as that of the above-mentioned two rocks.

It can be concluded from the above-mentioned deformation measurements under the true triaxial stress state that: The high anisotropic dilatancy occurs under high

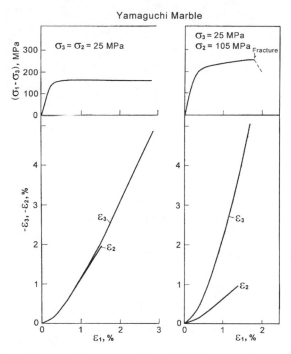

Figure 3.96.   Differential stress $(\sigma_1 - \sigma_3)$ and lateral strains ε_2 and ε_3 as functions of the axial strain ε_1 in Yamaguchi marble. Left: axi-symmetric stress state $(\sigma_2 = \sigma_3)$; right: the case of true triaxial compression.

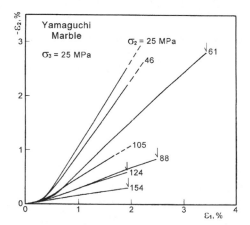

Figure 3.97.   Lateral strains ε_2 as functions of the axial strain ε_1 in Yamaguchi marble. Numerals for each curves are σ_2 values.

values of the intermediate principal stress σ_2. This anisotropic dilatancy may be explained by the opening of cracks perpendicular to the minimum principal stress σ_3. The stress level σ_1 at which dilatancy begins to occur increases with increase of σ_2 value.

3.3.i *Fracture of an inhomogeneous rock and an anisotropic rock*

In the above-mentioned experiments, homogeneous and isotropic rocks were consciously used, because the dispersion of experimental data is generally small in these rocks, and so they are suitable for accurate studies of the fundamental mechanical behavior of rocks. However, many rocks in the earth's crust are more or less inhomogeneous and anisotropic. In this section, preliminary experiments on complex rocks using the true triaxial compression apparatus are discussed.

3.3.i.1 *Inhomogeneous rock*

In an early experiment, the inhomogeneity of a rock sample was carefully considered (Mogi, 1959). In this experiment, a block of marble (45 cm × 35 cm × 4 cm) was used to obtain the effect of the stress rate on the fracture strength. In this case, it was important to minimize the dispersion of strength data.

For this purpose, 192 test specimens were cut off from the block of marble and numbered from 1 to 192 to indicate their original spatial position. The top figure in Fig. 3.98 shows the spatial distribution of the uniaxial fracture strength of each specimen obtained by the conventional method. The middle figure (2) shows the spatial distribution of the "Mean Strength" value of the test specimen, that is the average value of nine test specimens which consist of a central specimen and eight specimens surrounding it. The bottom figure (3) shows the spatial distribution of Young's modulus. Although the block of the marble is apparently homogeneous, the fracture strength and Young's modulus are not uniform and vary spatially systematically within the block, as shown in Fig. 3.98. The fracture strength (Mean Strength) is significantly different for different regions (A) and (B) indicated in the bottom figure. The average fracture strength is 81.8 MPa in region (A) and 78.2 MPa in region (B). The maximum value in region (A) is 83.5 MPa and the minimum value in region (B) is 74.8 MPa. The maximum difference in Mean Strength within the block of apparently homogeneous marble reached to 11%. By correction of this inhomogeneity, the effect of stress rate on fracture strength was obtained accurately. This research suggests that careful consideration of inhomogeneity of a rock sample may be useful in order to obtain accurate results in rock fracture problems.

Taking the above-mentioned result into consideration, a porous sandstone, which is a typical weak sedimentary rock, was tested using the true triaxial compression apparatus (Xu et al., 1980; Mogi, 1983). Although the mass of this rock seemed to be apparently homogeneous, the inhomogeneity was examined by the conventional triaxial compression test. For this purpose, 55 test specimens were cut off and were numbered from 1 to 55, as an indication of their original spatial position (Fig. 3.99). According to the conventional triaxial compression test ($\sigma_2 = \sigma_3$), there are two different types of deformation curves (A- and B-types) (Fig. 3.100). The fracture or yield stress of A-type specimen is generally higher than that of B-type specimen, and the fracture pattern is different in these two types, that is, a single shear fault trace can be seen in A-type specimen and the fracture pattern in B-type specimen is complex and suggest

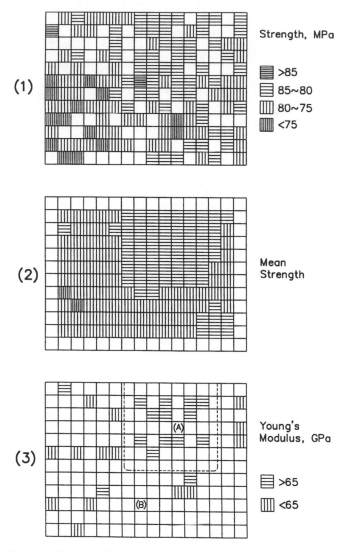

Figure 3.98. The distribution of fracture strength and Young's modulus in the sample of marble. "Mean Strength" value of the test specimen is the average value of nine test specimens which consist of the specimens around it and itself (Mogi, 1959).

rupture by compaction. This difference seems to be attributed to the difference in porosity.

In Fig. 3.99, the spatial distribution of specimens of A-type (solid circle) and B-type (open circle) is shown. These two types are not distributed at random, but they cluster in groups, that is, the mechanical behavior changes systematically and significantly within the block of the sandstone.

In Fig. 3.101, the results of true triaxial compression tests of the sandstone are shown. The maximum principal stress (σ_1) at rupture is plotted against the

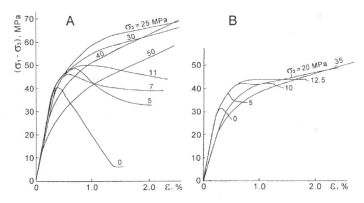

11	10	9	8	7	6	5	4	3	2	1
●	○	○	●	○	○	○	●	●	●	●

B   A

22	21	20	19	18	17	16	15	14	13	12
●	●	○	○	○	○	○	○	●	●	●

33	32	31	30	29	28	27	26	25	24	23
●	○	○	●	○	○	○	○	○	●	●

B

44	43	42	41	40	39	38	37	36	35	34
●	●	●	●	●	●	●	○	○	○	○

A

55	54	53	52	51	50	49	48	47	46	45
○	●	●	●	●	●	●	●	○	○	○

Figure 3.99.   55 test specimens were cut off from a block of a sandstone. Solid symbols: A-type (stronger); open symbols: B-type (weaker). (Xu et al., 1980).

Figure 3.100.   Stress-strain curves in A-type and B-type specimens (Xu et al., 1980).

intermediate principal stress (σ_2). In this case, since the peaks of the stress-strain curves are sometimes not clear, the stress at permanent strain $\varepsilon_p = 0.2\%$ or 0.4% was taken as the rupture stress (σ_1). However, the feature shown in Fig. 3.101 is not different for the two cases, and so the case of $\varepsilon_p = 0.2\%$ is shown in Fig. 3.101. The minimum principal stress (σ_3) is held at 10 MPa. Solid and open symbols indicate the data in region A and in region B, respectively. If all data are used without consideration of the above-mentioned inhomogeneity of this block, no significant result on the σ_2-effect can be obtained, because of great dispersion of data. However, if data in region A and region B are plotted individually, as shown in Fig. 3.101, these two groups of data do not overlap and their dispersion decreases markedly. From this result, the effect of σ_2 on the rupture strength (σ_1) can be recognized although the dispersion of data still remains. It is noted that the magnitude of the σ_2-effect in the weak sandstone is much smaller than that in the above-mentioned stronger rocks.

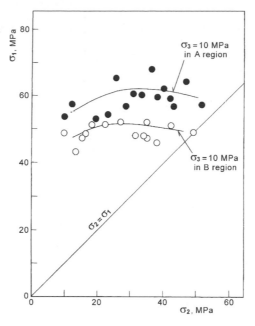

Figure 3.101.   Maximum principal stress at rupture or the stress at permanent strain $\varepsilon_p = 0.2\%$ (σ_1) as functions of σ_2. $\sigma_3 = 10$ MPa (Xu et al., 1980).

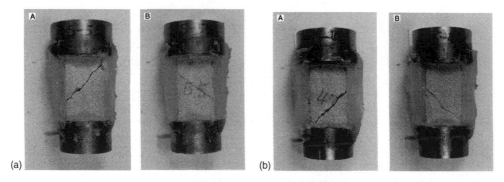

(a)   (b)

Figure 3.102.   (a) Sandstone No.155; (b) Sandstone No. 147.

Figures 3.102(a) and (b) show examples of the fracture pattern of specimens of sandstone. In these figures, an inclined fault trace is seen in A- and B-planes and the fault traces in A and B planes are symmetric. This result shows that the fracture occurred by shear faulting and the fault plane is roughly flat and parallel to the σ_2 direction, as observed in other rocks mentioned above.

Thus, the true triaxial compression experiment of an inhomogeneous weak sandstone was carried out by application of a careful method by which the effect of the non-uniform distribution of mechanical properties was reduced. The σ_2-effect of this weak sandstone is fundamentally similar to the other rocks. But the σ_2-effect on the

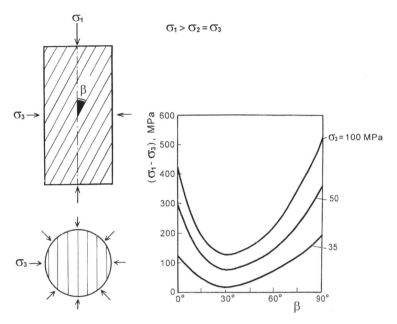

Figure 3.103.   Fracture strength ($\sigma_1 - \sigma_3$) of the anisotropic rocks as function of the angle (β) between the maximum principal stress (σ_1) and weak planes under the conventional triaxial compression (Donath, 1964).

strength is relatively small. However, this method for consideration of inhomogeneity may not always be applicable for other complex inhomogeneous rocks.

3.3.i.2   *Anisotropic rock*

In general, rocks or rock masses are more or less anisotropic, particularly, for example, the jointed rock masses or slates and schists. The effect of the stress states on failure behavior in anisotropic rocks has been studied by a number of researchers (e.g., Donath, 1961, 1964, 1972; McLamore and Gray, 1967) using the conventional triaxial compression test ($\sigma_2 = \sigma_3$). Among these, Donath exhaustively studied the strong influence of planar anisotropy on both fracture strength and fault angle. Figure 3.103 shows schematically his experimental method and the observed results (Donath, 1964). β is the angle between the maximum principal stress (σ_1) and weak planes, such as foliation planes. The right graph in Fig. 3.103 shows the differential stress ($\sigma_1 - \sigma_3$) at fracture as a functions of angle β for different confining pressure (Pc $= \sigma_2 = \sigma_3$). As can be seen in this figure (also in many other cases), the fracture strength of the anisotropic rock is lowest at around $\beta = 30°$. Since the fracture angles (θ) of isotropic rocks are nearly 30° in many cases, this result shows that fracture of anisotropic rocks is likely to occur when the angle β, showing the orientation of the weak planes of anisotropic rock, is nearly equal to the fracture angle (θ) of isotropic rocks with roughly similar solidity.

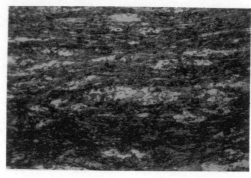

0.2 mm

Figure 3.104.   A photograph of preparate of Chichibu green schist cut normal to the foliation plane.

An experimental study of the effect of the more general triaxial stress state, that is, of the effect of the intermediate principal stress (σ_2), was carried out in 1977–79 by the above-mentioned true triaxial compression apparatus (Mogi et al., 1978; Mogi, 1979, 1983; Kwaśniewski and Mogi, 1990, 1996, 2000). A macroscopically homogeneous green crystalline schist with a distinct, dense foliation, from Chichibu mountain, west of Tokyo was selected for the experiment. Figure 3.104 shows a photograph of preparate of Chichibu green schist cut normal to the foliation plane.

3.3.i.2.1   Fracture stress

Figure 3.105 shows the relation between the maximum principal stress at fracture (σ_1) and the confining pressure ($\sigma_2 = \sigma_3$) of Chichibu green schist, obtained by the conventional triaxial compression test. The relation is shown for the two cases of $\beta = 30°$ (the direction of the lowest strength) and $\beta = 90°$ (the direction of the highest strength). Fracture strength increases linearly in both cases. The different symbols along the line for $\beta = 30°$ correspond to different angle ω which will be explained below.

By the true triaxial compression method, the measurements for the four cases: **I, II, III** and **IV**, which are shown in Fig. 3.106, were carried out. β is the angle between the σ_1 direction and the foliation planes, as mentioned above. ω is the angle between the σ_2 direction and the traces of planar foliations on the plane, which is normal to the σ_1 direction, as shown in Fig. 3.106. In this figure, orientation of three principal stresses (σ_1, σ_2 and σ_3) relative to the foliation planes is schematically shown.

The experimental results in the case **I** are shown in Figs. 3.107 and 3.108. In this case, it should be remarked that the orientation of the foliation plane nearly agrees with that of the fault plane at fracture (θ) of many isotropic rocks. It is particularly important that both the planes are parallel to the σ_2 direction.

Figure 3.107 shows the stress-strain curves in the direction of the maximum principal stress (σ_1) for different σ_2 values in case **I**. In this figure, the vertical axis is the

Table 3.14. Fracture stress of Chichibu green schist by conventional triaxial compression tests.

No.	σ_1 (MPa)	σ_2 (MPa)	σ_3 (MPa)
I–2-1	59	0	0
I–2-2	71	0	0
I–2-3	154	25	25
I–2-9	244	50	50
I–2-10	328	75	75
II-3-1	66	0	0
II-3-4	68	0	0
II-3-2	171	25	25
II-3-3	233	50	50
II-3-6	243	50	50
II-3-7	244	50	50
II-3-8	325	75	75
II-3-22	336	75	75
IV-1-1	147	0	0
IV-1-2	149	0	0
IV-1-14	294	25	25
IV-1-3	385	50	50
IV-1-4	501	75	75

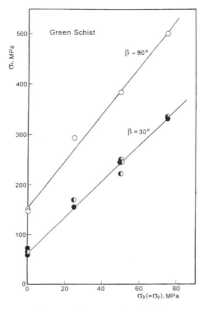

Figure 3.105. Relation between the maximum principal stress at fracture (σ_1) and the confining pressure ($\sigma_2 = \sigma_3$) of Chichibu green schist, obtained by the conventional triaxial test. (Kwaśniewski and Mogi, 1990).

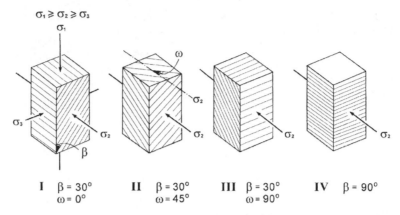

I $\beta = 30°$ II $\beta = 30°$ III $\beta = 30°$ IV $\beta = 90°$
    $\omega = 0°$ $\omega = 45°$ $\omega = 90°$

Figure 3.106.   Experiments for the four cases (**I, II, III** and **IV** of different orientations of foliation planes) by the true triaxial compression for green schist.

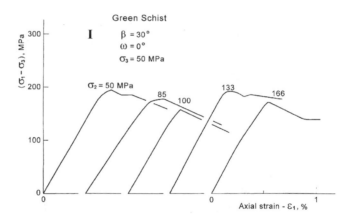

Figure 3.107.   Stress ($\sigma_1 - \sigma_3$)–strain curves for different σ_2 values in the case **I**, for green schist. Numericals for each curve are σ_2 values. $\sigma_3 = 50$ MPa.

differential stress, and the horizontal axis is the strain in the σ_1 direction. The minimum principal stress (σ_3) is held at the constant value (50 MPa) in this true triaxial compression test, for all cases **I–IV**. From this figure, it may be said that the stress-strain curves for different σ_2 values are nearly similar in case **I**.

In Fig. 3.108, the fracture stress σ_1 in case **I** (solid circle) is plotted against σ_2. Small open circles and a broken line show the relation between σ_1 and σ_3 obtained from the conventional triaxial compression tests. In case **I**, in which the foliation planes are parallel to the σ_2 direction and with the angle $\beta = 30°$, the fracture stress (σ_1) is nearly constant, that is, the σ_2-effect cannot be recognized in this case.

Figure 3.109 shows the relation between the maximum principal stress at fracture (σ_1) and the intermediate principal stress (σ_2) in case **II**. In this case, σ_1 increases

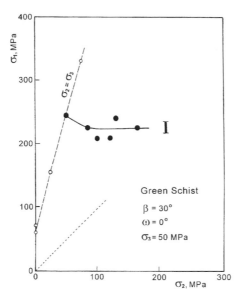

Figure 3.108.   Relation between fracture stress (σ_1) and σ_2 in the case **I**, for green schist (large solid circle). Small open circle: the case of $\sigma_2 = \sigma_3$.

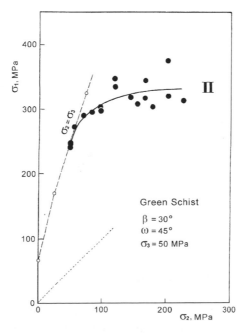

Figure 3.109.   Relation between fracture stress (σ_1) and σ_2 in the case **II**, for green schist (large solid circle). Small open circle: the case of $\sigma_2 = \sigma_3$.

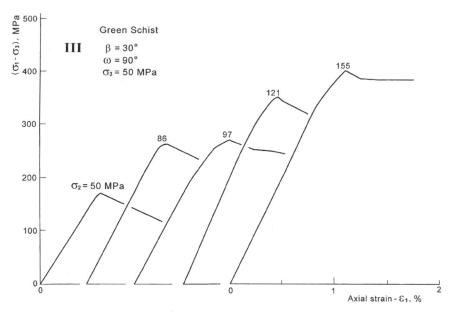

Figure 3.110.   Stress $(\sigma_1 - \sigma_3)$ – strain (ε_1) curves for different σ_2 values in the case **III**, for green schist. Numerals for each curve are σ_2 values. $\sigma_3 = 50$ MPa.

significantly with increasing σ_2 in the low σ_2 region, but the slope of the $\sigma_1 - \sigma_2$ curve gradually decreases with σ_2 until it reaches zero. However, it is noted that the dispersion of strength data is remarkable in the region of high σ_2 values. In this case, it may be concluded that the σ_2-effect is clearly recognized, but this effect is moderate.

Figure 3.110 shows examples of the stress-strain curves in the direction of the maximum principal stress (σ_1) in the true triaxial compression test in case **III**. Numerals for each curve are σ_2 values and σ_3 is 50 MPa. In this case, the slopes of these stress-strain curves are nearly similar, but the differential stress at peak of stress-strain curve markedly increases with σ_2. Figure 3.111 shows the fracture strength (σ_1) as a function of σ_2. In the case **III**, in which the σ_2 direction is perpendicular to the traces of foliation planes, the fracture stress (σ_1) quickly and linearly increases with the increasing σ_2. In this case, it is easy to understand that the increase of σ_2 strongly affects the fracture stress, because the frictional resistance along foliation planes increases effectively with an increase of σ_2.

Figure 3.112 shows examples of the stress-strain curves in the direction of the maximum principal stress (σ_1) in the true triaxial compression test in case **IV**. Numerals for each curve are σ_2 values and σ_3 is 50 MPa. The maximum differential stress increases significantly with σ_2. Figure 3.113 shows the fracture strength (σ_1) as a function of σ_2. In case **IV**, in which the σ_1 direction is perpendicular to the foliation planes $(\beta = 90°)$, σ_1 increases markedly with an increase of σ_2 in the initial stage, but the slope of the $\sigma_1 - \sigma_2$ curve gradually decreases with increasing σ_2. The σ_2-effect on the fracture stress in this case is very similar to that of the isotropic rocks, mentioned earlier. Since

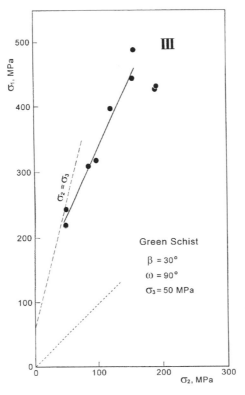

Figure 3.111.   Relation between fracture stress (σ_1) and σ_2, in the case **III**, for green schist (large solid circle). Broken line: $\sigma_1 - \sigma_2$ relation in the case of $\sigma_2 = \sigma_3$.

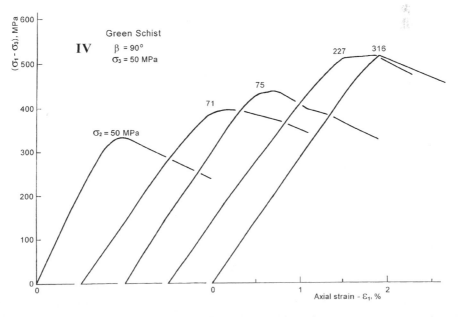

Figure 3.112.   Stress ($\sigma_1 - \sigma_3$) - strain (ε_1) curves for different σ_2 values in the case **IV**, for green schist. Numerals for each curve are σ_2 values. $\sigma_3 = 50\,\text{MPa}$.

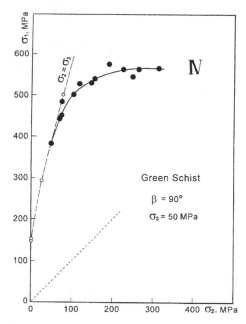

Figure 3.113.   Relation between fracture stress (σ_1) and σ_2 in the case **IV**, for green schist (large solid circle). Small open circle: the case of $\sigma_2 = \sigma_3$.

the foliation planes are perpendicular to the σ_1 direction, the existence of foliation planes probably does not have any important effect on rock failure characteristics.

The above-mentioned experimental results for the four cases (Figs. 3.108, 3.109, 3.111 and 3.113) of Chichibu green schist are summarized in Fig. 3.114. Differences of fracture strength and of the σ_2-effects for different orientations of the foliation planes can be clearly seen. The angle ω between the σ_2 direction and the trace of the planar foliation on the horizontal plane (when the σ_1 direction is vertical) is $0°$, $45°$ and $90°$ in case **I**, case **II** and case **III**, respectively. The effect of the intermediate principal stress σ_2 is negligible in case **I** (the σ_2 direction is parallel to the weak planes), moderate in case **II**, and very remarkable in case **III** (the σ_2 direction is perpendicular to the traces of the weak planes). In case **IV** ($\beta = 90°$), the fracture stress is the highest, because the weak planes do not have a significant influence on the failure behavior of the anisotropic rock. Although further experiments in more varied orientations of weak planes are needed to obtain a definite conclusion, this result seems to give an important suggestion for understanding of the problem "why σ_2 affects the failure behavior of isotropic rocks?".

Figure 3.115 shows a simple model to understand the σ_2-effect (Mogi, 1983). The left figure shows schematically the σ_2-effects for the three typical orientations of weak planes of an anisotropic rock which correspond to the above-mentioned cases **I**, **II** and **III**. The magnitude of the σ_2-effect is strongly dependent on the orientation of weak planes of an anisotropic rock. The right figure shows schematically a model which

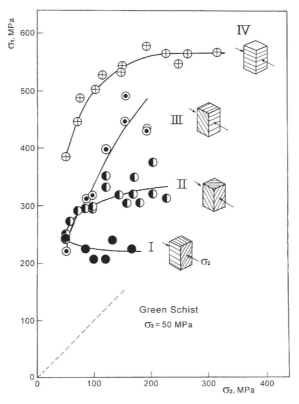

Figure 3.114.   Summary of $\sigma_1 - \sigma_2$ relation in the four different orientations of foliation planes for green schist.

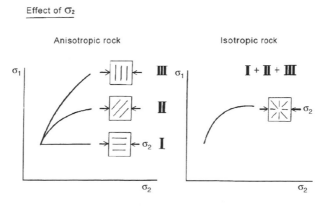

Figure 3.115.   A simple model to understand the σ_2 effect (Mogi, 1983). See text.

explains the σ_2-effect in isotropic rock, on the basis of the experimental results for an anisotropic rock, mentioned above. In isotropic rocks, small-scale cracks or grain boundaries with various orientations are probably distributed at random. The σ_2-effect in such rocks which contain many small-scale weak planes with various orientations, may be approximately represented by the average of the σ_2-effect of the three typical cases (**I**, **II** and **III**). This is shown schematically in the right graph in Fig. 3.115.

Experimental results obtained for an anisotropic rock are very interesting in themselves and further studies are still needed, as mentioned above. In addition, the results seem to give a key to clarify the fundamental problem of how the intermediate principal stress (σ_2) affects the failure strength of isotropic rocks, which is discussed preliminarily on the basis of limited data in this section.

Numerical data of fracture stress of Chichibu green schist obtained from true triaxial compression tests are shown in Table 3.15.

3.3.i.2.2   Strain and dilatancy

For measurements of strains in the σ_1 direction, the electric strain gages were used, in principles, for the initial stage preceding the failure of the specimen. However, since the strain is not uniform in anisotropic rocks, the reliability of this method is limited, particularly in the post- failure region. Therefore, the axial deformation of the specimens in the post-failure region was determined based on measurements of the movement of vertical piston in the pressure vessel. The measurements were taken by means of an LVDT installed outside the pressure vessel. The results were corrected for the elastic distortion of the piston, load cell and end pieces (Kwaśniewski and Mogi, 1990).

Figure 3.116 shows stress-strain curves including the post-failure region of Chichibu green schist in the σ_1 direction under true triaxial compression conditions in the four cases (**I**, **II**, **III** and **IV**). In cases **I**, **II** and **IV**, the stress drop is clearly observed, but the deformation in the post-failure region in case **III** is quite different in the cases of higher σ_2 values, in which the peak stress is not clear and a sudden stress drop does not occur.

In Fig. 3.117, the strain at the maximum point of the differential stress ($\sigma_1 - \sigma_3$) is plotted against ($\sigma_2 - \sigma_3$). In cases **I**, **II** and **IV**, the strain does not change significantly with an increase of σ_2, but the strain in case **III** increases markedly with an increase of σ_2. This marked σ_2-dependency corresponds to that of fracture stress shown in Fig. 3.111.

Using the method mentioned in the section 3.3.h.3, strains in the σ_1, σ_2 and σ_3 directions were measured in Chichibu green schist and the volumetric strain, including also dilatancy, was determined. Examples of these results in case **II** are shown in Figs. 3.118, 3.119(a), 3.119(b) and 3.119(c), and an example in case **IV** is shown in Fig. 3.120.

Figure 3.118 shows the case of $\sigma_2 = \sigma_3$. These curves are very similar to those of isotropic rocks, mentioned earlier. In this axi-symmetric stress state, curves of ε_2 and ε_3 are nearly similar. In the true triaxial compression state (Figs. 3.119 (a), 3.119 (b), 3.119 (c)), however, the strain (ε_3) in the direction of the minimum principal stress (σ_3) increases with the increase of the intermediate principal stress (σ_2). As can be seen in Fig. 3.119 (c), plot of differential stress versus ε_2 is completely linear and ε_3 increases markedly with the increase of ($\sigma_1 - \sigma_3$). The differential strain – volumetric strain

Table 3.15. Fracture stress of Chichibu green schist by true triaxial compression tests.

No.	σ_1 MPa	σ_2 MPa	σ_3 MPa
I $\beta = 30°$, $\omega = 0°$			
I-2-9	244	50	50
I-2-16	225	85	50
I-2-4	206	100	50
I-2-15	208	121	50
I-2-2	240	133	50
I-2-11	225	166	50
II $\beta = 30°$, $\omega = 45°$			
II-3-6	247	50	50
II-3-3	233	50	50
II-3-1	244	50	50
II-3-24	273	56	50
II-3-23	290	70	50
II-3-7	294	86	50
II-3-9	300	97	50
II-3-13	298	97	50
II-3-5	346	121	50
II-3-21	335	121	50
II-3-15	318	144	50
II-3-14	308	156	50
II-3-16	319	167	50
II-3-11	346	168	50
II-3-19	304	180	50
II-3-12	320	203	50
II-3-17	376	203	50
II-3-18	313	227	50
III $\beta = 30°$, $\omega = 90°$			
III-4-5	219	50	50
III-4-9	244	50	50
III-4-6	311	86	50
III-4-12	318	97	50
III-4-1	397	121	50
III-4-7	445	155	50
III-4-8	488	156	50
III-4-13	428	192	50
III-4-14	432	193	50
IV $\beta = 90°$			
IV-1-1	385	50	50
IV-1-17	446	71	50
IV-1-9	447	74	50
IV-1-16	486	75	50
IV-1-11	502	103	50
IV-1-5	528	115	50
IV-1-8	531	148	50
IV-1-13	540	156	50
IV-1-6	577	192	50
IV-1-15	563	227	50
IV-1-7	546	249	50
IV-1-10	562	262	50
IV-1-12	565	316	50

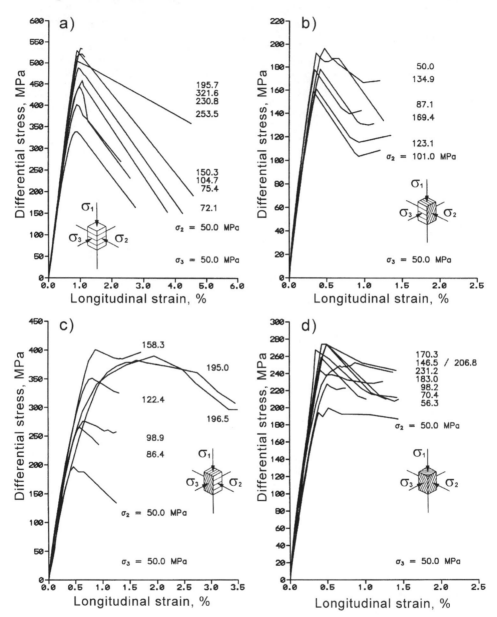

Figure 3.116.   Stress-strain curves including the post-failure region of green schist in the σ_1 direction under true triaxial compression conditions; (a) the case **IV**, (b) the case **I**, (c) the case **III**, (d) the case **II** (Kwaśniewski and Mogi, 1990).

($\Delta V/V_0$) characteristic deviates gradually from a straight line, that is, the inelastic volumetric strain (dilatancy) begins to increase at a certain stress level. This figure shows that the dilatancy is caused by the increase of the strain ε_3 in the direction of the minimum principal stress (σ_3).

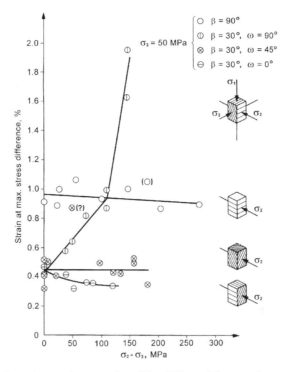

Figure 3.117. Strain at the maximum point of the differential stress $(\sigma_1 - \sigma_3)$ is plotted against $(\sigma_2 - \sigma_3)$ (Kwaśniewski and Mogi, 1990). See text.

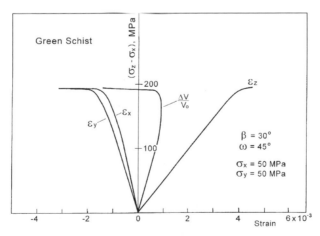

Figure 3.118. Three principal strains $(\varepsilon_X, \varepsilon_Y$ and $\varepsilon_Z)$ and volumetric strain $\Delta V/V_0$ as functions of differential stress $(\sigma_Z - \sigma_X)$. (X, Y and Z show the direction of the minimum principal stress σ_3, the direction of the intermediate principal stress σ_2 and that of the maximum principal stress σ_1, respectively.) This graph shows the result of the case **II** by the conventional triaxial compression test, $\sigma_2 = \sigma_3 = 50$ MPa.

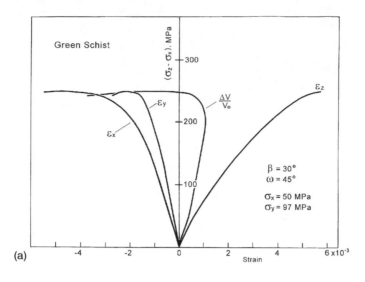

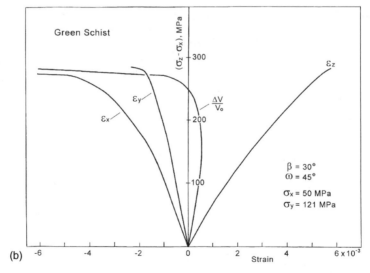

Figure 3.119. (a) Three principal strains (ε_X, ε_Y and ε_z) and volumetric strain $\Delta V/V_0$ as functions of differential stress ($\sigma_z - \sigma_X$) for green schist. X, Y and Z are the same as in Fig. 3.118. This graph shows the result of the case **II** by the true triaxial compression test, $\sigma_2 = 97\,\text{MPa}$ and $\sigma_3 = 50\,\text{MPa}$. (b) Three principal strains and volumetric strain curves in the case **II**, $\sigma_2 = 121\,\text{MPa}$ and $\sigma_3 = 50\,\text{MPa}$, for green schist. (c) Three principal strains and volumetric strain curves in the case **II**, $\sigma_2 = 168\,\text{MPa}$ and $\sigma_3 = 50\,\text{MPa}$, for green schist.

Figure 3.120 shows an example of results in case **IV**. In this case, stress-strain curves are similar to those typical of isotropic rocks. Since the foliation planes in this case are perpendicular to the σ_1-direction, the anisotropy of rocks does not have a significant effect on the failure behavior.

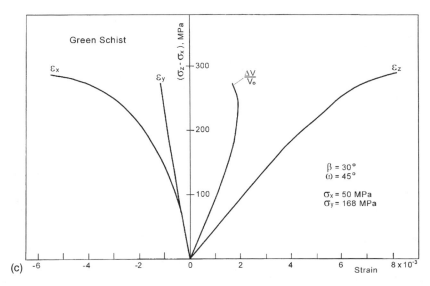

Figure 3.119. (*Continued*)

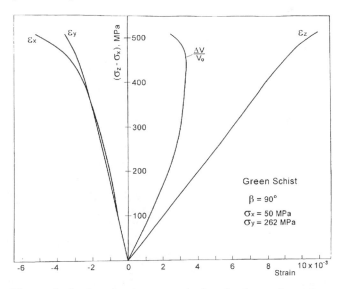

Figure 3.120. Three principal strains (ε_X, ε_Y and ε_Z) and volumetric strain curves in the case **IV**, $\sigma_2 = 262\,\text{MPa}$ and $\sigma_3 = 50\,\text{MPa}$, for green schist.

    As mentioned earlier, since the deformation of anisotropic rocks is not homogeneous in the test specimens, it should be noticed that the accuracy of these strain measurements is limited.

3.3.i.2.3   Fracture pattern

Fracture patterns of anisotropic rocks are strongly influenced by orientations of weak planes under true triaxial compression (Kwaśniewski and Mogi, 1990, 1996, 2000). Figure 3.121(a) shows an example of results in case **I** in which the foliation planes

180   *Experimental rock mechanics*

Figure 3.121.   Fracture pattern of Chichibu green schist. (Kwaśniewski and Mogi, 1990).
(a)–(f); (a) Chichibu green schist, No. **I**-2-4. $\sigma_1 = 206\,\text{MPa}$, $\sigma_2 = 100\,\text{MPa}$, $\sigma_3 = 50\,\text{MPa}$;
(b) Chichibu green schist, No. **II**-3-1. $\sigma_1 = 244\,\text{MPa}$, $\sigma_2 = 50\,\text{MPa}$, $\sigma_3 = 50\,\text{MPa}$; (c) Chichibu
green schist, No.**II**-3-9. $\sigma_1 = 300\,\text{MPa}$, $\sigma_2 = 97\,\text{MPa}$, $\sigma_3 = 50\,\text{MPa}$; (d) Chichibu green schist,
No. **II**-3-11. $\sigma_1 = 346\,\text{MPa}$, $\sigma_2 = 168\,\text{MPa}$, $\sigma_3 = 50\,\text{MPa}$; (e) Chichibu green schist, No.
IV-1-13. $\sigma_1 = 540\,\text{MPa}$, $\sigma_2 = 156\,\text{MPa}$, $\sigma_3 = 50\,\text{MPa}$; (f) Chichibu green schist, No. **IV**-1-10.
$\sigma_1 = 562\,\text{MPa}$, $\sigma_2 = 262\,\text{MPa}$, $\sigma_3 = 50\,\text{MPa}$.

are parallel to the σ_2 direction. Plane A and plane B are the two planes perpendicular to the σ_2 direction on both sides of a rectangular parallelepiped rock specimen, as shown in Fig. 3.77. In this case, a single, linear, inclined fault can be clearly seen, and the traces of faults in planes A and B are symmetric. This figure shows that the main fault is very flat and parallel to the σ_2 direction. It is most probable that the faulting occurred along one of the foliation planes of the schist.

Figure 3.121 (b) shows the fracture patterns of the test specimen of case **II** under the axi-symmetric stress state. Traces of 4 or 5 inclined slip planes are seen and the traces of slip planes on planes A and B are nearly symmetric. These slips probably occurred along the foliation planes.

Figure 3.121 (c) and (d) show the fracture patterns of the test specimens of case **II** under true triaxial compression. The σ_2 values are different in these tests. In these figures, fracture patterns in plane A and plane B are not completely symmetric and are complex to some degree. In this case, the fault planes are not parallel to the σ_2-direction.

In case **III**, in which the strike of the inclined foliation planes is perpendicular to the σ_2 direction ($\omega = 90°$), the fault planes are not parallel to the σ_2 direction, but to the σ_3 direction. The strong influence of σ_2 on the fracture strength of Chichibu green schist may be attributed to the large difference in the directions of the strikes of the foliation planes and σ_2. In this case, the frictional strength along the planar foliations increases with an increase of σ_2, because the normal stress on the foliation planes increases with an increase of σ_2. In this experiment, the magnitude of σ_2 was limited. However, if experiments are carried out for larger σ_2 values, the fracture pattern may change.

Figure 3.121 (e) and (f) show the fracture patterns in case **IV** in which the foliation planes arc perpendicular to the σ_1 direction. In Fig. 3.121 (e) and (f), the fracture patterns are somewhat complex. This may be due to some inhomogeneity of the test specimen, but the main fault is nearly similar to that of the above-mentioned case of homogeneous rocks.

Although the orientation of the foliation planes of Chichibu green schist and the magnitude of stresses are limited, the preliminary study shows how strongly the intermediate principal stress affects the ultimate strength, strain, and fracture patterns. Further studies on various anisotropic rocks and different orientations of weak planes under true triaxial compression conditions are essentially important.

3.3.j   *Other recent experiments*

Xu Dongjun and Geng Naiguang carried out true triaxial compression experiments at the Earthquake Research Institute, University of Tokyo at the end of 1970's. Then, Xu Dongjun constructed a Mogi-type true triaxial compression apparatus in China (Xu, 1980). His group carried out a number of true triaxial compression experiments using this machine (Xu and Geng, 1984, 1985). In particular, they studied the σ_2-effect in the region of high σ_2 values including $\sigma_2 = \sigma_1$, and reported that σ_1 increased with an

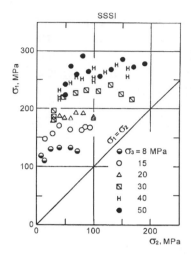

Figure 3.122.   Fracture strength (σ_1) as a function of the intermediate principal stress (σ_2) in Shirahama sandstone (Takahashi and Koide, 1989).

increase of σ_2 and reached the maximum value, and then gradually decreased. They also reported on the effect of loading paths.

In the late 1970's Koide constructed a large Mogi-type triaxial compression apparatus at the Geological Survey of Japan. This apparatus is a nearly similar version (not the same) of the apparatus developed by Mogi at the Earthquake Research Institute of the University of Tokyo in the late 1960's (Mogi, 1969, 1970a). It is equipped with an electric-hydraulic servo-controlled system of sample deformation and can accommodate larger samples (maximum 5.0 × 5.0 × 10.0 cm). Takahashi and Koide (1989) carried out true triaxial compression experiments on three sandstones, a shale and a marble. The size of the samples used in these experiments was 3.5 × 3.5 × 7.0 cm. Figures 3.122–3.124 show some of results of their experiments, where the maximum principal stress at fracture (σ_1) is plotted against the intermediate principal stress (σ_2). The general pattern of the results agrees with those revealed by Mogi and discussed above. They obtained a series of axial differential stress versus strain curves of Yamaguchi marble for different σ_2 values under $\sigma_3 = 20$ MPa, and pointed out that the onset of dilatancy increased with an increase of σ_2. They also proposed a modified effective shear strain energy criterion. In Takahashi and Koide's experiments, the applied stresses and the fracture strength of tested rocks were relatively low and the σ_2-effect is also small, but the results may be applicable to engineering problems.

Haimson and Chang (2000) designed and fabricated a true triaxial testing system suitable for testing strong rocks at the University of Wisconsin, U.S.A., which emulates Mogi's original design with significant simplification, and reported results of true triaxial tests on Westerly granite. Their experimental results on the relationship between σ_1 at strength failure and σ_2 are shown in Fig. 3.125(a). Chang and Haimson (2000 a, b) carried out true triaxial compression tests on a KTB amphibolite, which is

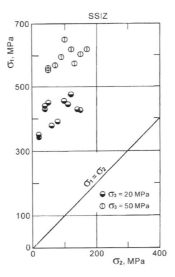

Figure 3.123.   Fracture strength (σ_1) as a function of the intermediate principal stress (σ_2) in Izumi sandstone (Takahashi and Koide, 1989).

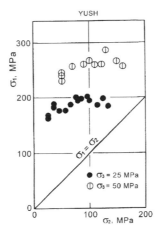

Figure 3.124.   Fracture strength (σ_1) as a function of the intermediate principal stress (σ_2) in Yuubari shale (Takahashi and Koide, 1989).

a rock extracted from a superdeep scientific borehole (depth 9.1 km) drilled under the auspices of the German Continental Deep Drilling Program (KTB) in southeastern Germany. The tested rock samples came from a depth of 6.4 km. The $\sigma_1 - \sigma_2$ relations for different σ_3 values obtained from the true triaxial compression tests are shown in Fig. 3.125(b). These results are similar to the results of Manazuru andesite, Inada granite and Orikabe monzonite (Figs. 3.48, 3.51. 3.52) reported by Mogi (1970b).

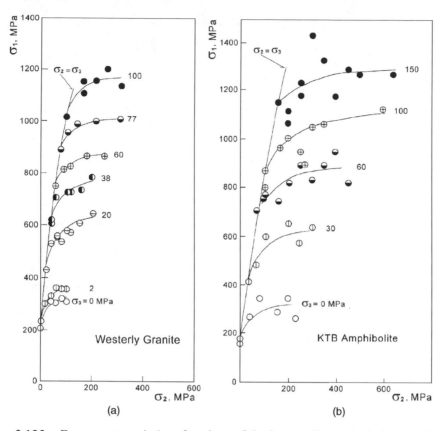

Figure 3.125.   Fracture stress (σ_1) as functions of the intermediate principal stress (σ_2) for hard rocks. (a) Westerly granite (Haimson and Chang, 2000). (b) KTB amphibolite (Chang and Haimson, 2000b). Scales of σ_1-and σ_2-axes are same.

Chang and Haimson (2000b) discussed the following fracture strength criterion proposed by Mogi (1971a, b):

$$\tau_{\text{oct}} = f[(\sigma_1 + \sigma_3)/2] \tag{3.23}$$

and presented it in the following form:

$$\tau_{\text{oct}} = 1.51[(\sigma_1 + \sigma_3)/2]^{0.89} \quad \text{for Westerly granite} \tag{3.24}$$

$$\text{and} \quad \tau_{\text{oct}} = 1.73[(\sigma_1 + \sigma_3)/2]^{0.86} \quad \text{for KTB amphibolite} \tag{3.25}$$

In their papers, they pointed out that the onset of dilatancy increases with an increase of σ_2, as was first shown by Mogi (1977) and Takahashi and Koide (1989).

In any discussion of papers by Haimson and Chang (2000) and Chang and Haimson (2000 b), the following problems should be noted:

(1) In Fig. 3.125 (a) and (b), the relations between σ_1 and σ_2 for $\sigma_3 \approx 0$ are shown. However, as discussed in Chapter 1, the end effect in the case of $\sigma_3 \approx 0$ should be

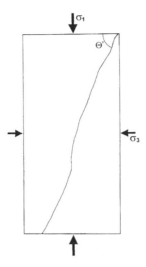

Figure 3.126.   Fracture pattern of a KTB amphibolite specimen after a true triaxial test by Chang and Haimson (2000b). Only a main fault in their photograph is reproduced by the present author.

carefully considered in such hard rocks, because the stress distribution in the test specimen is not homogeneous under very low confining pressure. Therefore, the observed strength values for $\sigma_3 \approx 0$ may be not precise.

(2) In Fig. 7 of the paper by Chang and Haimson (2000b), the trace of the main fracture is not linear and significantly curved, as can be seen in the fracture trace reproduced in Fig. 3.126. In the Table 2 in their paper, the fracture angle (θ') in this case is given equal to 66°. The determination of fracture angle from such non-linear fault traces is open to question.

Furthermore, the following studies on the σ_2-effect should be remarked. In 1973, Atkinson and Ko reported a fluid cushion multiaxial cell for testing cubical rock specimens. (Atkinson and Ko, 1973). Michelis (1985) proposed a true triaxial cell using fluid cushion and reported some experimental results, but his results are open to question because of low experimental reliability.

3.3.k   *Future problems*

The stress states in the earth's crust are generally complex. Therefore, studies on the deformation and fracture behavior of rocks and rock masses under the general stress states are very important. Since the pioneering work by von Kármán (1911), a large number of experiments on rock deformation and strength have been carried out under axi-symmetric compressive stress conditions ($\sigma_2 = \sigma_3$). However, as mentioned above, the number of experiments on rock samples under true triaxial stress states ($\sigma_1 \geq \sigma_2 \geq \sigma_3$) is very limited. The author of this book designed a satisfactory true triaxial compression apparatus and carried out many tests on samples of several rocks, and published several papers (Mogi, 1970–1983), as mentioned in this chapter.

The results of these studies were presented at the International Upper Mantle Symposium at Flagstaff, AZ, USA in 1970 (Mogi, 1971b), the 15th General Assembly of International Union of Geodesy and Geophysics in Moscow, the Soviet Union in 1971 (Mogi, 1972b), the 3rd International Congress of the ISRM in Denver, CO, USA in 1974 (Mogi, 1974b), and the 4th International Congress of the ISRM in Montreux, Switzerland in 1979 (Mogi, 1979) as an invited speaker. However, I became busy in the national project for earthquake prediction in Japan, and was installed as a vice-chairman (April 1981–March 1991) and chairman (April 1991–March 2001) of the Coordinating Committee for Earthquake Prediction, Japan. This is the main reason why I could not continue my research work in this field of experimental rock mechanics. During this period, my colleagues Kwaśniewski (Poland) and Xu (China) have continued true triaxial experiments with their groups. In particular, Kwaśniewski, Takahashi and Li have been working together since 2000 at the Center for Deep Geological Environments, the National Institute of Advanced Industrial Science and Technology (AIST) in Tsukuba. Using Koide and Takahashi's true triaxial compression apparatus, they have been studying the effect of both σ_2 and σ_3 on the dilatant behavior and permanent volumetric strain in clastic, sedimentary rocks (Kwaśniewski et al., 2002, 2003).

In recent years, the importance of the σ_2-effect on the mechanical and hydro-mechanical behavior of rocks has been seen in a new light. In my opinion, the following issues are important and should be addressed in future studies:

(1) Effect of mineralogical, structural and textural features of different rock materials,
(2) Effect of magnitude of both σ_2 and σ_3
(3) Effect of stress orientation relative to the planes of stratification, foliation, etc. (anisotropy)
(4) Effect of stress or strain rate
(5) Effect of loading paths
(6) Effect of water (dry or wet, pore pressure)
(7) Effect of temperature

It should be stressed, that it is of the utmost importance that true triaxial compression experiments should be carried out with great care and extreme accuracy.

REFERENCES

Adams, F. D. and J. T. Nicolson. (1901). An experimental investigation into the flow of marble, Royal Soc. London Philos. Trans. Ser. A, **195**, 597–637.

Al-Ajimi, A. M. and R. W. Zimmerman. (2005). Relation between the Mogi and the Coulomb failure criteria. Int. J. Rock Mech. Min. Sci., **42**, 431–439.

Atkinson, R. H. and Hon-Yim Ko. (1973). A fluid cushion, multiaxial cell for testing cubical rock specimens. Int. J. Rock Mech. Min. Sci. & Geomech. Abstr., **10**, 351–361.

Böker, R. (1915). Die Mechanik der bleibenden Formänderung in Kristallinisch aufgebauten Körpern. Ver. Dtsch. Ing. Mitt. Forsch., **175**, 1–51.

Brace, W. F. (1964). Brittle fracture of rocks. In: State of stress in the Earth's crust. Judd, W. R. (ed.), Elsevier, New York, 111–174.

Brace, W. F., B. W. Paulding, and C. H. Scholz. (1966). Dilatancy in the fracture of crystalline rocks. J. Geophys. Res., **71**, 3939–3953.

Bresler, B. and K. S. Pister. (1957). Failure of plain concrete under combined stresses. Trans. Am. Soc. Civ. Eng., **122**, 1049–1068.

Chang, C. and B. Haimson. (2000a). Rock strength determination using a new true triaxial loading apparatus, and the inadequacy of Mohr-type failure criteria. In: Pacific Rocks 2000, Proc. 4th North Am. Rock Mech. Symp., Seattle, J. Girard, M. Leibman, C. Breeds & T. Doe (eds.). Balkema, Rotterdam, 1321–1327.

Chang, C. and B. Haimson. (2000b). True triaxial strength and deformability of the German Continental Deep Drilling Program (KTB) deep hole amphibolite. J. Geophys. Res., **105**, 18999–19013.

Colmenares, L. B. and M. D. Zoback. (2002). A statistical evaluation of intact rock failure criteria constrained by poly axial test data for five different rocks. Int. J. Rock Mech. Min. Sci., **39**, 695–729.

Donath, F. A. (1961). Experimental study of shear failure in anisotropic rocks. Bull. Geol. Soc. Am., **72**, 985–990.

Donath, F. A. (1964). Strength variation and deformational behavior in anisotropic rock. In : State of stress in the Earth's crust. Judd W. R. (ed.). New York, American Elsevier, 281–297.

Donath, F. A. (1972). Faulting across discontinuities in anisotropic rock. In : Stability of Rock Slopes. Cording, E. J. (ed.). Proc. 13th Symp. Rock Mech., Urbana, Ill., Aug/Sept 1971. New York : Am. Soc. Civ. Eng., 753–772.

Fjær, E. and H. Ruistuen. (2002). Impact of the intermediate principal stress on the strength of heterogeneous rock. J. Geophys. Res., **107**, No. B2, ECV 3, 1–9.

Haimson, B. and C. Chang. (2000). A new true triaxial cell for testing mechanical properties of rock, and its use to determine rock strength and deformability of Westerly granite. Int. J. Rock Mech. Min. Sci., **37**, 285–296.

Handin, J. and R. V. Hager. (1957). Experimental deformation of sedimentary rocks under confining pressure: tests at room temperature on dry samples. Bull. Am. Assoc. Petrol. Geol., **41**, 1–50.

Handin, J., D. V. Higgs and J. K. O'Brien. (1960). Torsion of Yule marble under confining pressure. In: Rock Deformation. Griggs, D., Handin, J. (eds.). Geol. Soc. Am. Memoir **79**, 245–274.

Handin, J., H. C. Heard and J. N. Magouirk. (1967). Effects of the intermediate principal stress on the failure of limestone, dolomite, and glass at different temperatures and strain rates. J. Geophys. Res., **72**, 611–640.

Heard, H. C. (1960). Transition from brittle fracture to ductile flow in Solenhofen limestone as a function of temperature, confining pressure, and interstitial fluid pressure. In: Rock Deformation. Griggs, D., Handin, J. (eds.). Geol. Soc. Am. Memoir **79**, 193–226.

Hojem, J. P. M. and N. G. W. Cook. (1968). The design and construction of a triaxial and polyaxial cell for testing rock specimens. S. Afr. Mech. Eng., **18**, 57–61.

Hoskins, E. R. (1969). The failure of thick-walled hollow cylinders of isotropic rock. Int. J. Rock Mech. Min. Sci., **6**, 99–125.

Jaeger, J. C. and E. R. Hoskins. (1966). Rock failure under the confined Brazilian test. J. Geophys. Res., **71**, 2651–2659.

Kármán, T. von. (1911). Festigkeitsversuche unter allseitigem. Druck. Z. Ver. Dtsch. Ing. **55**, 1749–1757.

Kwaśniewski, M. A. and K. Mogi. (1990). Effect of the intermediate principal stress on the failure of a foliated anisotropic rock. In: Proc. Int. Conf. Mech. Jointed and Faulted Rock, H. P. Rossmanith (ed.). Rotterdam, Balkema, 407–416.

Kwaśniewski, M. A. and K. Mogi. (1996). Faulting of a foliated rock in a general triaxial field of compressive stresses. In: Tectonophysics of Mining Areas (ed. A. Idziak). Wyd. Uniwersytetu Śląskiego, Katowice, 209–232.

Kwaśniewski, M. A. and K. Mogi. (2000). Faulting in an anisotropic, schistose rock under general triaxial compression. In: Pacific Rocks 2000, Proc. 4th North Am. Rock Mech. Symp., Seattle, J. Girard, M. Liebman, C. Breeds & T. Doe (eds.). Balkema, Rotterdam, 737–746.

Kwaśniewski, M. A., M. Takahashi and X. Li. (2002). On the dilatant behavior of sandstone under true triaxial compression conditions (Extended abst.). In: Mining Geophys. (ed. W. W. Zuberek), Pub. Inst. Geophys., Polish Acad. Sci., M-24 (340), 321–324.

Kwaśniewski, M. A., M. Takahashi and X. Li. (2003). Volume changes in sandstone under true triaxial compression conditions. In: Technology Roadmap for Rock Mechanics, Proc. 10th Congr. Int. Soc. Rock Mech., Vol. 1, 683–688.

Lode, W. (1926). Proc. Int. Congress Appl. Mech. 2nd, Zurich.

Matsushima, S. (1960). On the deformation and fracture of granite under high confining pressure. Bull. Disaster Prevention Res. Inst. Kyoto Univ., **36**, 11–20.

McClintock, F. A. and J. B. Walsh. (1962). Friction on Griffith cracks in rocks under pressure. Proc. 4th Natl.Congr. Appl. Mech., Am. Soc. Mech. Eng., New York, N. Y., 1015–1021.

McLamore, R. and K. E. Gray. (1967). The mechanical behavior of anisotropic sedimentary rocks. J. Engineering for Industry (Trans. Am. Soc. Mech. Eng. Ser. B), **89**, 62–73.

Michelis, P. (1985). A true triaxial cell for low and high pressure experiments. Int. J. Rock Mech. Min. Sci. & Geomech. Abstr., **22**, No. 3, 183–188.

Mjachkin, V., W. Brace, G. Sobolev and J. Dieterich. (1975). Two models of earthquake forerunners. Pure Appl. Geophys., **113**, 169–181.

Mogi, K. (1959). Experimental study of deformation and fracture of marble (1). On the fluctuation of compressive strength of marble and relation to the rate of stress application. Bull. Earthq. Res. Inst., Tokyo Univ., **37**, 155–170.

Mogi, K. (1966). Some precise measurements of fracture strength of rocks under uniform compressive stress. Felsmechanik und Ingeneiurgeologie, **IV/1**, 41–55.

Mogi, K. (1967). Effect of the intermediate principal stress on rock failure. J. Geophys. Res., **72**, 5117–5131.

Mogi, K. (1969). On a new triaxial compression test of rocks. In: Abstr. 1969 Meeting Seismol. Soc. Japan, 3.

Mogi, K. (1970a) Effect of the triaxial stress system on rock failure. Rock Mech. in Japan. ISRM National Group of Japan, **1**, 53–55.

Mogi, K. (1970b). Effect of the combined stresses on rock failure – Studies by a new triaxial compression technique (1). Bull. Earthq. Res. Inst., Tokyo Univ., (Received July 30, 1970, but unpublished) pp. 42. (see Appendix).

Mogi, K. (1971a). Effect of the triaxial stress system on the failure of dolomite and limestone. Tectonophysics, **11**, 111–127.

Mogi, K. (1971b). Fracture and flow of rocks under high triaxial compression. J. Geophys. Res., **76**, 1255–1269.

Mogi, K. (1971c). Failure criteria of rocks (Study by a new triaxial compression technique), J. Soc. Material Sci. Japan, **20**, No. 209, 143–150. (in Japanese with English Abstr.).

Mogi, K. (1972a). Effect of the triaxial stress system on fracture and flow of rocks. Phys. Earth Planetary Interiors., **5**, 318–324.

Mogi, K. (1972b). Fracture and flow of rocks. Tectonophysics, **13**, 541–568.

Mogi, K. (1974). Rock fracture and earthquake prediction. J. Soc. Materials Japan, **23**, No. 248, 320–331. (in Japanese with English Abstr.).

Mogi, K. (1977). Dilatancy of rocks under general triaxial stress states with special reference to earthquake precursors. J. Physics Earth, **25**, Suppl. S203–S217.

Mogi, K., K. Igarashi and H. Mochizuki. (1978). Deformation and fracture of rocks under general triaxial stress states – anisotropic dilatancy. J. Soc. Material Sci. Japan, **27**, No. 293, 148–154. (in Japanese with English Abstr.).

Mogi, K., M. A. Kwaśniewski and H. Mochizuki. (1978). Fracture of anisotropic rocks under general triaxial compression. In: Abstr., Seismol. Soc. of Japan Bull., No.1, p. 25. (in Japanese with English Abstr.).

Mogi, K. (1979). Flow and fracture of rocks under general triaxial compression. Proc. 4th Congress of the ISRM (Int. Soc. Rock Mech.), Montreux, Vol. 3, 123–130. Balkema, Rotterdam.

Mogi, K. (1983). Rheology of rocks under true triaxial compressive stress. Research Progress Report of a Grant-in Aid for Scientific Research from the Ministry of Education, Science and Culture, p. 46. (in Japanese).

Murrell, S. A. F. (1965). The effect of triaxial stress system on strength of rocks at atmospheric temperatures. Geophys. J. R. Astron. Soc., **10**, 231–281.

Nadai, A. (1950). Theory of Flow and Fracture of Solids. **1**, 2nd ed., New York, McGraw-Hill, pp. 572.

Niwa, Y., S. Kobayashi and W. Koyanagi. (1967). Failure criterion on light-weight aggregate concrete subjected to triaxial compression. Mem. Fac. Eng. Kyoto Univ. **29**, 119–131.

Nur, A. (1972). Dilatancy, pore fluids, and premonitory variations ts/tp travel times. Bull. Seismol. Soc. Am., **62**, 1217–1222.

Paterson, M. S. (1967). Effect of pressure on stress-strain properties of materials. Geophys. J. R. Astron. Soc., **14**, 13–17.

Paterson, M. S. (1978). Experimental Rock Deformation – The Brittle Field. Springer-Verlag, Berlin, pp. 254.

Paterson, M. S. and Wong T-f. (2005). Experimental Rock Deformation – The Brittle Field, Second edition. Springer, Berlin, pp. 347.

Robertson, E. C. (1955). Experimental study of the strength of rocks. Bull. Geol. Soc. Am. **66**, 1275–1314.

Sakurai, S. and S. Serata. (1967). Mechanical properties of rock salt under three dimensional loading conditions. Proc. Japan Congr. Test. Mater. **10**, 139–142.

Scholz, C. H., L. R. Sykes and Y. P. Aggarwal. (1973). Earthquake prediction: A physical basis. Science, **181**, 803–809.

Takahashi, M. and H. Koide. (1989). Effect of the intermediate principal stress on strength and deformation behavior of sedimentary rocks at the depth shallower than 2000 m. In: V. Maury and D. Fourmaintraux (eds.), Rock at Great Depth, Vol. 1, 19–26. Balkema, Rotterdam/Brookfield.

Wawersik, W. R., L. W. Carson, D. J. Holcomb and R. J. Williams. (1997). New method for true-triaxial rock testing. Int. J. Rock Mech. Min. Sci., **34**, Paper no. 330.

Wiebols, G. A. and N. G. W. Cook. (1968). An energy criterion for the strength of rock in polyaxial compression. Int. J. Rock Mech. Min. Sci., **5**, 529–549.

Xu, D. J., W. Liu, H. Mochizuki and K. Mogi. (1980). Mechanical behavior of soft sandstone under true triaxial compression. In: Abstr. 1980 Meeting Seismol. Soc. Japan.

Xu, D. J. and N. G. Geng. (1984). Rock rupture and earthquake caused by change of the intermediate principal stress. Acta Seismologica, **6**, (2), 159–166. (in Chinese).

Xu, D. J and N. G. Geng. (1985). The variation law of rock strength with increase of intermediate principal stress. Acta Mechanics Solida Sinica, **7**, 72–80. (in Chinese, English Abstr.).

APPENDIX

The text presented below is the first part of the original, unpublished manuscript by Kiyoo Mogi on the True Triaxial Experiment. This manuscript was the first systematic text on the effect of the intermediate principal stress on rock failure and included the complete experimental results that were obtained in 1970. It was received on July 30, 1970 by the Editorial Board of the Bulletin of the Earthquake Research Institute, University of Tokyo, Japan. Just after submission of the paper, the author was invited as a speaker for the International Upper Mantle Symposium in Flagstaff, U.S.A. in 1970. Convinced that the experiment by the True Triaxial Compression test should be reported at this symposium, the author proposed to stop the publication of the paper by the Earthquake Research Institute. His lecture at the Upper Mantle Symposium was very well received and it was subsequently published in the Journal of Geophysical Research (Mogi, 1971). The published text however contained only a summary of the essentials described in the original, unpublished manuscript. As this original paper forms the basis of Chapter 3 of this book and has historical value for the field, the abstract, contents and the part of the introduction of this text are included in this Appendix.

(1)

Effect of the combined stresses on rock failure — Study by a new
triaxial compression technique — (1)

By KIYOO MOGI

Earthquake Research Institute.

(Read Oct. 28, 1969; May 26, 1970. — Received July 30, 1970.)

Abstract

Failure under conditions of triaxial stresses, in which two of the
principal stresses are equal, have been intensely studied, but few investigations of the effect of the more general combined stress system have been
made because of the experimental difficulties of achieving conditions of
true homogeneous triaxial stress. A new triaxial apparatus has made possible the study of failure of brittle rocks under general triaxial stress
system in which all three principal stresses are different. By this method,
the effects of the stress system to fracture and yield strength, the stress
drop at faulting and the ductility in several rocks were experimentally
studied. The fracture strength increases not only with the increase of the
least compression σ_3, but also with the increase of the intermediate compression σ_2, except for very high σ_2. The yield strength also increases
with σ_3 and σ_2, although its σ_3 and σ_2 dependencies are different from those
of fracture strength. The influences of σ_2 and σ_3 on the ductility and the
stress drop, however, are just the opposite.

The following macroscopic criteria for fracture and yielding which were
obtained by generalization of the von Mises theory, correlates the present
results satisfactorily:

$$(2)$$

$$\tau_{oct} = f_1(\sigma_1 + \sigma_3) \qquad \text{for fracture,}$$

$$\tau_{oct} = f_2(\sigma_1 + \sigma_2 + \sigma_3) \qquad \text{for yielding.}$$

The physical interpretations of these criteria are as follows: failure will occur when the stored energy of distortion reaches a critical value dependent on the effective mean normal pressure which is $(\sigma_1 + \sigma_3)/2$ for fracture and $(\sigma_1 + \sigma_2 + \sigma_3)/3$ for yielding.

Contents

Introduction

Part I New triaxial compression technique

    1. New design of triaxial apparatus

    2. Sample design

    3. Experimental procedure

Part II Strength and ductility of rocks under triaxial compression

    1. Dunham dolomite

    2. Solenhofen limestone

    3. Yamaguchi marble

    4. Mizuho trachyte

    5. Manazuru andesite

    6. Inada granite

    7. Orikabe monzonite

    8. Conclusion

(3)

Part **III** Macroscopic criteria for fracture and yielding of rocks

   1. Previous studies

   2. Fracture criterion — generalized von Mises' criterion of fracture

   3. Yield criterion — generalized von Mises' criterion of yielding

   4. Conclusion

Introduction

    Failure behaviour of rocks has been intensely studied under triaxial compression, in which two of the principal stresses are equal. However, owing to experimental difficulties, few investigations of rock failure under general triaxial stress systems, in which all three principal stresses are different, have been made. In this paper, a new triaxial compression apparatus is described in which each stress can be controlled independently and the effects of general triaxial stresses on failure behaviour for several rocks have been investigated.

    A state of stress that is necessary to produce failure in an element can be described by the principal stresses σ_1, σ_2 and σ_3. In this paper, compressive stress is taken positive and $\sigma_1 > \sigma_2 > \sigma_3$. In principal coordinates, the point $(\sigma_1, \sigma_2, \sigma_3)$ representing different states of stress just necessary failure might form a surface

$$\sigma_1 = f(\sigma_2, \sigma_3) \qquad (1)$$

for a given material at a constant temperature, constant strain rate, etc. A fundamental problem of rock mechanics is the study of the shape of this surface for various rocks. For ductile metals, it is now well established that the limiting surface of yielding is obtained from the von Mises' criterion (Nadai, 1950). For more brittle materials such as rocks and concrete,

II
Acoustic Emission (AE)

CHAPTER 4

AE Activity

4.1 INTRODUCTION

The term acoustic emission (AE) is widely used to denote the phenomenon in which a material or structure emits elastic waves of shock type and sometimes of continuous type caused by the sudden occurrence of fractures or frictional sliding along discontinuous surfaces. This phenomenon occurs in a variety of cases, and so it has been studied in various fields, such as material science, civil engineering, mining engineering, seismology. Therefore, at the early stages the phenomenon was studied independently in different fields and different terms were used in different fields, that is, for example, acoustic emission in metal engineering and microseismic activity in mining engineering. The present author used the term elastic shock in his early papers. Since the 1970s, the term acoustic emission has been widely used. In this chapter, some subjects of laboratory studies on acoustic emission phenomena in rocks and rock-like materials will be mentioned mainly on the basis of the author's experiments.

Probably the first laboratory measurement of acoustic emission phenomenon caused by the fracturing of solid materials was carried out by a Japanese seismologist Kishinouye (1937). He measured high frequency elastic waves caused by fractures resulting from the bending of a beam-shaped wood specimen for the purpose of simulating the temporal variation of the activity of 1930 Ito earthquake swarm in Japan. In the 1940s, Obert and Duvall (e.g. 1945) carried out laboratory experiments on acoustic emission phenomenon as a part of their research on the problems of mine design and rock burst prevention. In metal engineering, Kaiser (1953) found an important property of acoustic emission, which later was termed the Kaiser effect.

In seismology, several researchers reported laboratory experiments on acoustic emission phenomenon in relation to some problems of earthquakes (e.g. Kōmura, 1955; Shamina, 1956; Vinogradov, 1959; Scholz, 1968a). A series of systematic laboratory experiments on acoustic emission phenomenon by the present author for various cases (Fig. 4.1) began in 1959 (e.g. Mogi, 1962a, 1968) based on the viewpoint that earthquakes were large-scale acoustic emission events. Since the late 1960s, acoustic emission study has been promoted markedly and it has developed as a new field in rock physics (e.g. JSNDI, 2000).

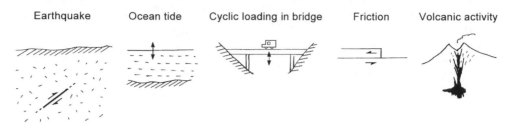

Figure 4.1.   Various cases in AE generation.

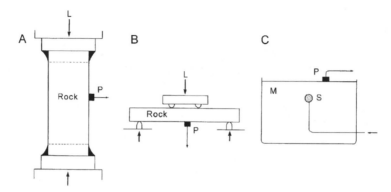

Figure 4.2.   Schematic figures of simple three different experimental systems of AE measurements in laboratory. A: case of uniaxial compression; B: case of bending; C: case of stress application by an inner pressure source (S). P: transducer for AE measurement.

4.2   AE ACTIVITY UNDER SOME SIMPLE LOADINGS

The temporal variation of AE activity of rocks and rock-like materials, which are heterogeneous brittle materials, has been measured using different experimental arrangements (Mogi, 1962ab, 1963a, 1968). Figure 4.2 shows three different cases: **A**, **B** and **C**. In case **A**, the uniaxial uniform compressive load was applied to a cylindrical rock specimen, whose design is explained in detail in Chapter 1 (1.1). In case **B**, a bending load was applied to a beam-shaped specimen using a four-point loading method. In case **C**, stress was generated in a semi-infinite brittle medium by an inner stress source. P in Fig. 4.2 denotes a transducer for measurement of AE events which is connected to an amplifier-recorder system. The case **C** shows a case of a simple model experiment of earthquakes.

Figure 4.3 shows the two most fundamental cases of stress application. The left figure shows a case where the stress increases at a constant rate up to the main fracture. The temporal variations of strain and the frequency of AE events in this case are also shown schematically. The right figure shows the case where stress is applied suddenly and kept constant up to the main fracture. The temporal variation of strain and the frequency of AE events in this case are also shown schematically.

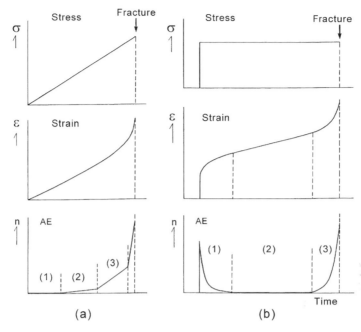

Figure 4.3.   Two fundamental cases of stress application (a) and (b), and temporal variations of strain (ε) and the frequency (n) of AE events in these cases.

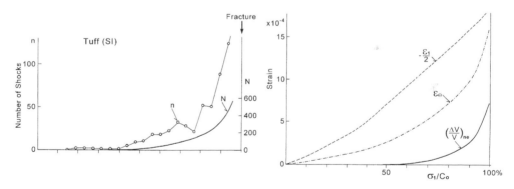

Figure 4.4.   Temporal variations of number of AE events and axial strain (ε_1), lateral strain (ε_θ) and non-elastic volumetric strain $(\Delta V/V)_{ne}$.

Figure 4.4 shows the experimental result for Saku-ishi tuff under uniform uniaxial compression conditions (case **A**). AE activity and strains are shown as a function of the compressive stress σ_1 normalized by the uniaxial compressive strength (C_o). n and N are the number of AE events and their accumulated number, respectively. ε_1 and ε_θ are the axial strain and the lateral strain, respectively. $(\Delta V/V)_{ne}$ is the non-linear volumetric strain. The lateral strain and AE activity increase markedly up to the main fracture. Particularly, it is noted that the accumulated number of AE events and the non-linear volumetric strain (dilatancy) increase simultaneously before fracture, as

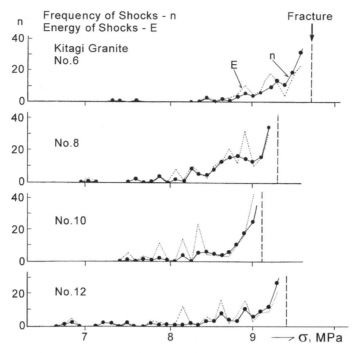

Figure 4.5.   AE activity as function of increasing bending stress for Kitagi granite specimens.

also mentioned by Scholz (1968a). This result shows that AE events occurred due to micro-fracturing of the rock material.

Figure 4.5 shows the experimental results for Kitagi granite under bending stress which was increased at a constant rate (case **B**). The frequency curves of AE events are shown as functions of tensile stress. The relative value of the elastic wave energy (E) of successive AE events per time unit was calculated as follows:

$$E = K \Sigma \alpha^2 \qquad (4.1)$$

where K is a constant and α is the maximum trace amplitude of an elastic wave of shock type (AE). The temporal variations of energy (E) per time unit are shown by dotted curves in Fig. 4.5. According to these results, AE events begin to occur at the middle stress level and increase markedly with increasing stress, particularly just before the main fracture. This pattern is very similar to that observed during compression of cylindrical rock specimens.

Figure 4.6 shows the temporal variations of AE activities of Kitagi granite specimens under a constant bending stress condition (case **B**). Under the continuous application of constant stress, rock specimens deform gradually and rupture after stress has been applied for some time, as shown schematically in Fig. 4.3 (b). The time interval between the beginning of loading and the occurrence of fracture is called "fracturing time". For the same rock material, the fracturing time varies considerably

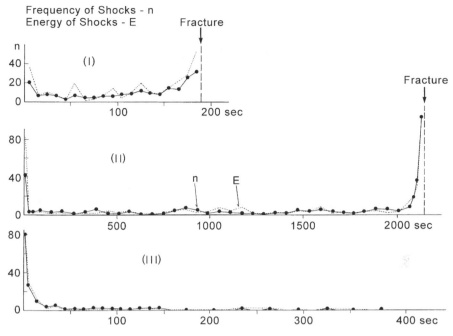

Figure 4.6. Temporal variation of AE activity of Kitagi granite specimens under a constant bending stress.

for each specimen. This is a noticeable feature of the brittle fracture of solid materials, because brittle fracture is a stochastic process. The three different patterns of AE activity shown in Fig. 4.6 are typical examples in which the fracturing time is different. As shown schematically in Fig. 4.3 (b), under a constant stress, the processes of deformation and AE activity consist of (1) the transient creep stage, (2) the steady creep stage and (3) the tertiary creep stage. AE activity in the transient creep stage is shown in Fig. 4.7. The temporal change of AE source location in the steady creep and the tertiary creep stages will be discussed in detail in the following chapter.

As can be seen in Fig. 4.6, the frequency of AE events n(t) decreases monotonically after the stress application in the transient creep stage. Figure 4.7 (a) shows the accumulated frequency curve of AE events for case III in Fig. 4.6. The accumulated frequency $N(t)$ is defined as

$$N(t) = \int_t^\infty n(t)dt \tag{4.2}$$

The relation between log $N(t)$ and t seems to be nearly linear in a semi-logarithmic scale graph, except for the initial stage. Therefore, the frequency $n(t)$ can be approximately expressed as

$$n(t) = n_0 e^{-\mu t} \tag{4.3}$$

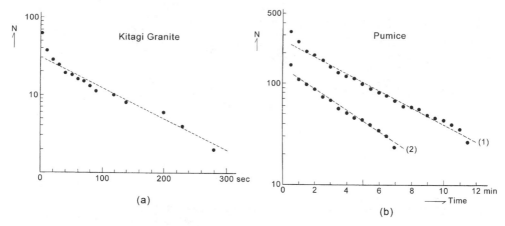

Figure 4.7.   Accumulated frequency $N(t)$ of AE events under constant stress as functions of time. (a) Kitagi granite; (b) Volcanic pumice.

where n_0 and μ are constants. This means that the frequency of AE events of Kitagi granite under a constant stress decreases exponentially with time.

In Fig. 4.7 (b), the experimental result for a volcanic pumice is shown. Volcanic pumice particles which are brittle and highly porous were compressed under constant load conditions. The accumulated frequency curves of AE events in the granular pumice under a constant stress for two experiments are shown in Fig. 4.7 (b). The relation between log $N(t)$ and t is nearly linear, which means that the frequency of AE events decreases exponentially with time.

These results indicate that AE events in brittle heterogeneous rocks in the transient creep stage occur with a constant transition probability. This decay curve of AE activity is similar to the decay curve of radioactivity.

The fracturing time of Inada granite and Komikage granite specimens was measured by application of constant bending load to the beam-shaped specimens. The frequency histogram for fracturing time of Inada granite specimens and Komikage granite specimens under constant bending load are shown in Fig. 4.8. The relation between the accumulated frequency (N) of fracturing time defined by Equation (4.2) and time for the two granites are linear in semi-logarithmic graphs as shown in Fig. 4.9. Therefore, it can be concluded that the rock fracture under constant stress occurs with the constant transition probability μ, and μ depends on the applied stress σ. To obtain the quantitative relation between μ and σ, the mean fracturing time t_m was measured under various constant stresses, because the mean fracturing time has the following relation to μ:

$$t_m = 1/\mu \tag{4.4}$$

Experimental results plotted in Fig. 4.10 show the following relationship between the applied stress (σ) and the mean fracturing time (t_m) or the transition probability of

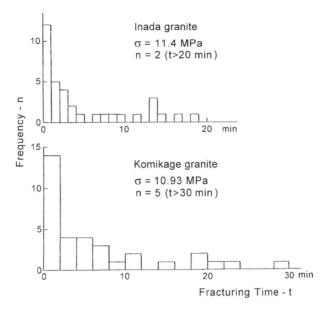

Figure 4.8.   Frequency histograms for fracturing time (t) of granite specimens under the constant stress (σ).

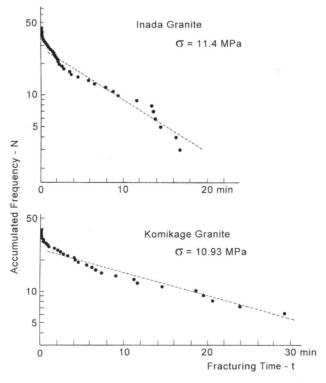

Figure 4.9.   Accumulated frequency curve of fracturing time (t) of granite specimens under the constant stress (σ).

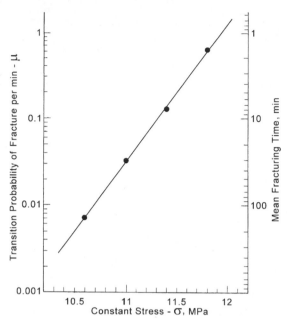

Figure 4.10.   Relation between the applied stress (σ) and the mean fracturing time (t_m) or the transition probability of fracture per min (μ) for Inada granite.

fracture (μ) of Inada granite:

$$\mu = Ae^{B\sigma} \tag{4.5}$$

where A and B are constants depending on the strength of materials. Although the above-mentioned discussion is particularly applicable to the main fracture of rock specimens, this theory can be applied to AE events which occur due to micro-fracturing of brittle heterogeneous rock specimens (Mogi, 1962a).

Thus, AE activity is strongly dependent on the stress level, as shown above. In addition to the stress level, the degree of heterogeneity and brittleness (or ductility) also play an important role. Figure 4.11 shows AE activity of rock and rock-like materials with different degrees of heterogeneity under linearly increasing stresses (the stress is compressive in pumice specimens and bending in other materials). AE activity markedly increases with increasing heterogeneity, as can be seen clearly in this figure. It has been the present author's opinion since the 1960s that heterogeneity is a key factor in the fracture process of brittle materials. This will be discussed further in the following section.

AE events in rock specimens occur mainly due to small or large-scale brittle fractures. Therefore, it is apparent that AE activity is dependent on the degree of brittleness. Figure 4.12 shows curves of inelastic strain (ε_n) and the accumulated energy of AE events (W_s) in granite and andesite specimens under increasing bending load conditions. Although granites and andesites are crystalline igneous rocks, the relation between ε_n and W_s is quite different in these two rocks. In the case of andesites the inelastic strain ε_n increases markedly and AE activity is considerably

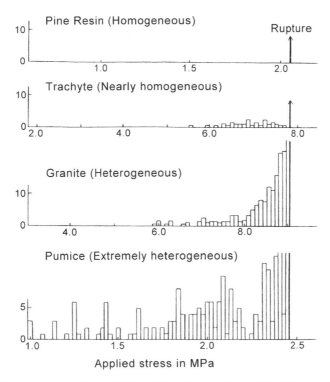

Figure 4.11. Temporal variation of AE activity of materials with different degree of hetero-geneity under linearly increasing stress. Pumice: compressive stress; other specimens: bending stress.

low before the fracture, thus the main part of inelastic strain may be attributed to ductile deformation. On the other hand, in the case of granites the increase in inelastic strain is not so noticeable and AE activity is very high, thus the main part of inelastic strain may be attributed to the occurrence of many micro-cracks. Thus, AE activity is strongly dependent upon the degree of brittleness of rocks.

4.3   THREE PATTERNS OF AE ACTIVITY

Until the late 1950th, it has been generally believed that the mode of successive occurrence of earthquakes is quite complicated and that there are no simple regularities, except for Ōmori's formula for aftershock frequency curve (Ōmori, 1894). However, according to a systematic investigation of earthquake sequences, the patterns of earthquake sequence were classified into the following three types (Fig. 4.13) (Mogi, 1963b).

(1) The first type is a large earthquake without foreshocks, but frequently followed by many aftershocks. Most large tectonic earthquakes are of this type.
(2) The second type are those where foreshocks occur prior to the principal earthquake and numerous aftershocks follow it.

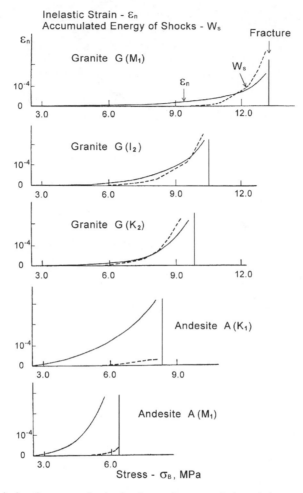

Figure 4.12.   Relation between the inelastic strain curve (ε_n) and the curve of energy (W_s) emitted as AE events. The inelastic strain is estimated from the difference between the total strain and the elastic strain which can be measured.

(3) The third type is the earthquake swarm where the number and the magnitude of the earthquakes increase gradually with time and decrease after some duration. There is no single predominant principal earthquake in the swarm.

Examples of the three types of earthquake sequences are shown in Fig. 4.14 (see examples in Mogi (1985)). In order to study the mechanism of these various types of earthquake sequences, patterns of sequences of model earthquakes were experimentally investigated from the viewpoint that natural earthquakes are large scale AE events. As a simple and general model of a shallow earthquake in laboratory, the fracture of a semi-infinite medium caused by an inner stress source was studied. Experimental arrangement of the model experiment is shown schematically in

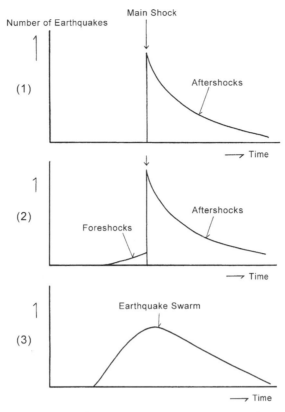

Figure 4.13.　Three types of earthquake sequences.

Fig. 4.15. Most of natural earthquakes occur by shear faulting of limited length by external stresses. However, stress was applied by inner stress sources as a simple model for the purpose of the experiment. The inner stress source was buried at some depth in the semi-infinite body and its pressure was increased gradually over a period of time. The fracture occurred between the stress source and the surface of the medium. This is somewhat similar to that of natural shallow earthquakes. AE events accompanying the brittle fracture were measured in the cases of different mediums and stress sources. Various patterns of sequences of AE events were obtained.

　　Experimental results are shown in Fig. 4.16–Fig. 4.18, and summarized in Fig. 4.19. The sequences of AE events in the model experiments can also be classified into three fundamental types like the natural earthquake sequences. The difference between the types is due to the structural state of the medium as well as to the distribution of the applied stress. That is:

(1) The first type occurs in a homogeneous medium under nearly uniform applied stress.

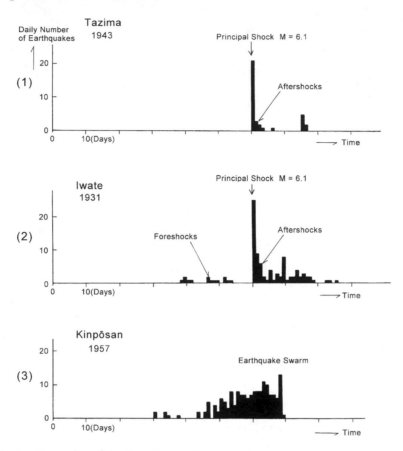

Figure 4.14.   Examples of the three types of earthquake sequences.

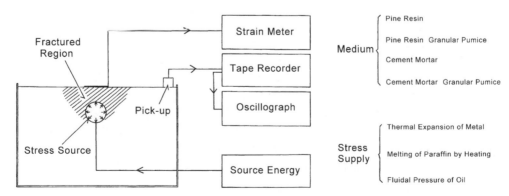

Figure 4.15.   Experimental arrangement for a simple model experiment for study of the process of earthquake generation.

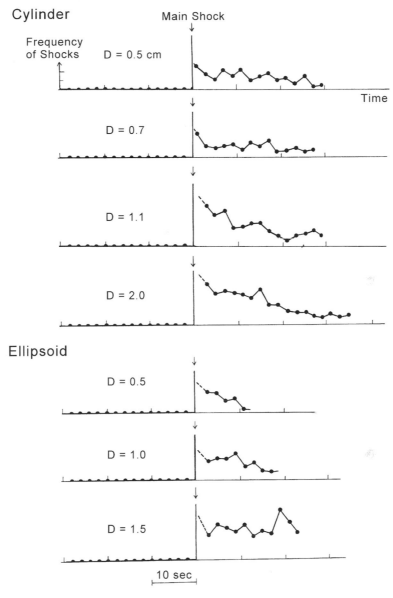

Figure 4.16. Temporal variation of number of AE events accompanying fractures of a semi-infinite homogeneous medium caused by an inner stress source of round shape. D: depth of stress source.

(2) The second type occurs when the structure and/or the space distribution of the applied stress are not uniform. In this case, small AE events occur prior to the main event and many small AE events follow it.

(3) The third type, the swarm type, occurs in a highly nonuniform medium and/or by the application of a concentrated stress.

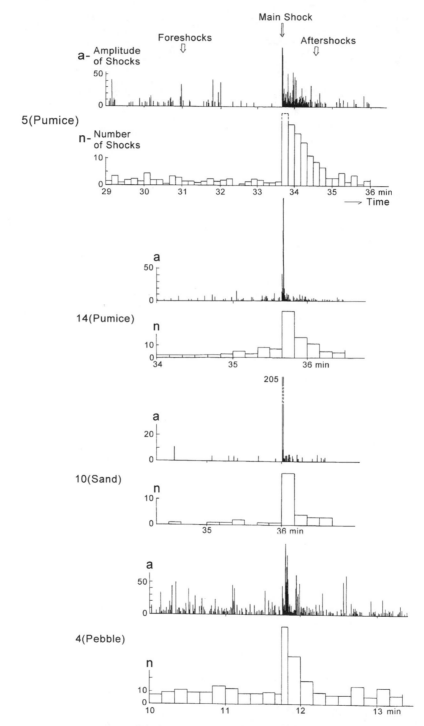

Figure 4.17. Temporal variation of AE activity (amplitude and number) accompanying fractures of a semi-infinite heterogeneous medium caused by an inner stress source.

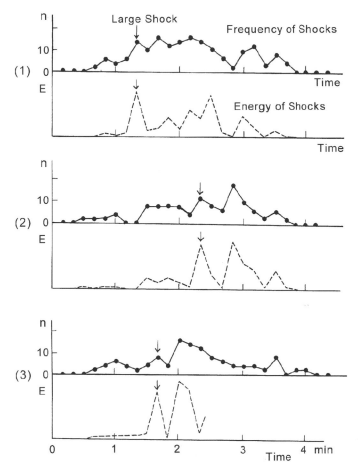

Figure 4.18. Temporal variation of AE activity (number of events and energy) accompanying fractures of a semi-infinite medium with high degree of heterogeneity by an inner stress source.

Thus, the difference between one type and another is attributed to the difference in the degree of stress concentration in the medium.

From the above-mentioned experimental result, we can deduce that the type of earthquake sequence depends on the structural state of the earth's crust and the space distribution of the applied stresses. In tectonic earthquakes, since the stress generally seems to be applied uniformly, the difference between one type and another is attributed to the differences in the structural state. At shallow regions, the structural state of the earth's crust may depend upon the degree of fracturing. It is therefore deduced that foreshocks take place in moderately heterogeneous regions and non-volcanic active swarms take place in highly fractured regions.

In and around the active volcanoes, mechanical structures at shallow depth are generally very complex and stresses are applied nonuniformly, for example, by intrusion of magma. Actually the swarm-type shock sequences occur frequently in this case (Mogi, 1963c).

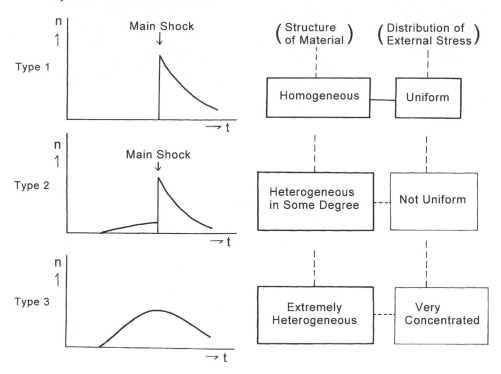

Figure 4.19.   Three types of AE sequences accompanying fractures of a semi-infinite medium and their relation to the structure and the stress distribution.

Figure 4.20 shows the location of the regions where Type 2 and Type 3 sequences shown in Fig. 4.19 frequently occur (Mogi, 1963b). This result indicates that fore-shocks and earthquake swarms do not occur accidentally, but occur in groups at certain places. Aftershocks also occur with a similar tendency. Figure 4.21 shows an example of the typical Type 2 sequence, that is, the Kita-izu earthquake (M 7.3) which occurred on 26th November, 1930 in the Fuji volcanic zone, central Honshū, Japan, that was preceded by a noticeable foreshock sequence.

According to the above-mentioned discussion, the type of earthquake sequence may be attributed to the degree of fracturing of the earth's crust. By the same token, the mechanical structure of the earth's crust can be deduced from the locations of different types of earthquake sequences. This is shown in Fig. 4.22. Region *1*, where earthquake swarms and foreshocks frequently occur, is the highly fractured region. Region *2*, where foreshocks frequently occur, is the next most mechanically disturbed region. Only aftershocks occur in Region *3*, the moderately disturbed region. In the mechanically undisturbed region, Region *4*, mainly isolated earthquakes occur. This mechanical structure agrees well with the recent geotectonic structure deduced from geological and geophysical data. The space distribution of the fractured regions agree with that of Quaternary and Neogene Tertiary volcanic regions and Tertiary folded regions (Fig. 4.23). This result could be explained because the volcanic zones are

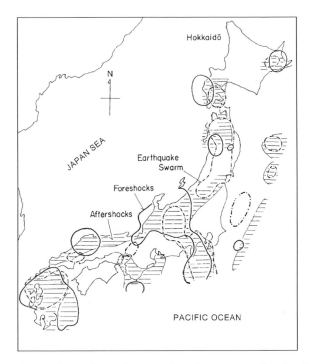

Figure 4.20.   Location of the regions where earthquake swarms, foreshocks and aftershocks frequently occur, in and near Japan.

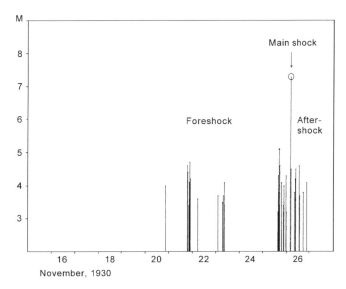

Figure 4.21.   Sequence of foreshocks, main shock and aftershocks in the case of the 1930 Kita-izu (central Honshū, Japan) earthquake of M 7.3. (Data from JMA Catalogue).

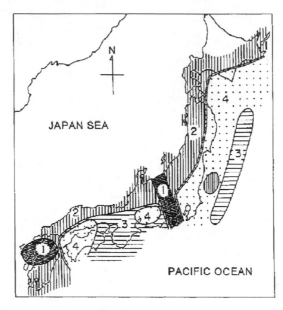

Figure 4.22.   Mechanical structure of the earth's crust in and near Japan deduced from the locations of different types of earthquake sequences. The number in each region indicates the degree of fracturing.

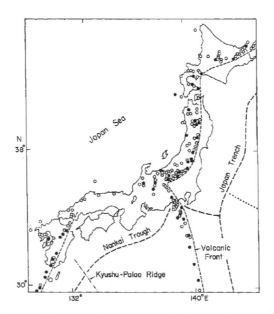

Figure 4.23.   Locations of Quaternary volcanoes and a trench in and near Japan. Closed circle: active volcano; open circle: non-active volcano; dotted line in the Pacific Ocean: zone of sea mounts.

thought to be fractured zones of the earth's crust. The most fractured region at the central part of Honshū corresponds to the very important tectonic break which crosses Honshū and separates the Japanese Islands into a southwest and northeast tectonic division. Some of the most active volcanoes in Japan are located in this zone.

REFERENCES

Japanese Soc. Non-destructive Inspection (JSNDI). (2004). Progress in Acoustic Emission, **10**, JSNDI, Tokyo.

Kaiser, J. (1953). Erkenntnisse und Folgerungen aus der Messung von Gerauschen bei Zugbeanspruchung von Metallischen Werkstoffen. Archiv für das Eisenhüttenwesen, **24**, 43–45.

Kishinouye, F. (1937). Frequency-distribution of the Ito earthquake swarm of 1930. Bull. Earthquake Res. Inst., Tokyo Univ., **15**, 785–827.

Kōmura, S. (1955). On relation between the Ishimoto-Iida statistical formula and some crack phenomena. J. Seismol. Soc. Japan, **8**, 80–83. (in Japanese)

Mogi, K. (1962a). Study of elastic shocks caused by the fracture of heterogeneous materials and its relation to earthquake phenomena. Bull. Earthquake Res. Inst., Tokyo Univ., **40**, 125–173.

Mogi, K. (1962b). The fracture of a semi-infinite body caused by an inner stress origin and its relation to earthquake phenomena (1). Bull. Earthquake Res. Inst., Tokyo Univ., **40**, 815–829.

Mogi, K. (1963a). The fracture of a semi-infinite body caused by an inner stress origin and its relation to the earthquake phenomena (2). Bull. Earthquake Res. Inst., Tokyo Univ., **41**, 595–614.

Mogi, K. (1963b). Some discussions of aftershocks, foreshocks and earthquake swarms — the fracture of a semi-infinite body caused by an inner stress origin and its relation to the earthquake phenomena (3). Bull. Earthquake Res. Inst., Tokyo Univ., **41**, 615 658.

Mogi, K. (1963c). Experimental study on the mechanism of the earthquake occurrence of volcanic origin. Bull. Volcanol., Ser.II, **26**, 197–208.

Mogi, K. (1964). Laboratory experiments on fracture of rocks with different degrees of heterogeneity. US-Japan Conf. Research related to Earthquake Prediction Problems, 57–58.

Mogi, K. (1967). Earthquakes and fractures. Tectonophysics, **5**, 35–55.

Mogi, K. (1968). Source locations of elastic shocks in the fracturing process in rocks (1). Bull. Earthquake Res. Inst., Tokyo Univ., **46**, 1103–1125.

Mogi, K. (1985). Earthquake Prediction. Academic Press, Tokyo, pp. 355.

Obert, L. and Duvall, W. (1945). Microseismic method of predicting rock failure in underground mining. Part II. Laboratory experiments. U. S. Bur. of Min., Rep. Invest. 3803, p 14.

Ōmori, F. (1894). On the aftershocks of earthquakes. Report Imperial Earthquake Investigation Committee, **2**, 103–139. (in Japanese)

Scholz, C. H. (1968). Microfracturing and the inelastic deformation of rock in compression. J. Geophys. Res., **73**, 1417–1432.

Shamina, O. G. (1956). Laboratory investigation of process of crack propagation. Bull. Acad. Sci. USSR., No. 5, 513–518.

Vinogradov, S. D. (1959). On the distribution of the number of fractures in dependence on the energy liberated by the destruction of rocks. Izvestya Akad. Nauk, SSSR, Ser. Geofiz., 1850–1852.

CHAPTER 5

Source location of AE

5.1   INTRODUCTION

Source location studies of AE events in rock samples are interesting and important aspects of rock physics. The first study was carried out by Scholz (1968) on cylindrical specimens of Westerly granite under uniaxial compressive stress conditions. Then, Mogi (1968) carried out source location measurements on beam-shaped specimens of various rocks under bending load conditions. While reviewing these early studies Hardy (1972) stated the following:

"Although Scholz's studies represent a pioneering effort in laboratory scale acoustic emission source location in geologic materials, with the limited studies carried out, and the questionable location accuracy ($\Delta\chi i \approx 0.1$ in.), the results should be considered with caution. Recent comments by Harding (1970) and Mogi (1968) further reinforce this opinion. Probably the most outstanding laboratory source location study carried out to date with respect to geologic materials is that by Mogi (1968). In order to provide a simple geometry, Mogi utilized beam shaped specimens deformed under four-point loading. During these studies Mogi investigated specimens of three marbles, three granites, a trachyte and an andesite."

The present author wrote in the introduction to his paper (Mogi, 1968): "Recently, Scholz (1968) tried to determine source locations of elastic shocks prior to rupture of a granite specimen, by use of initial motions of S-waves. He remarked a clustering of micro-fracturing on the eventual fault plane. In his study, however, a total of twenty-two shocks only were located and results were accompanied by appreciable errors, so that detailed features in source locations in the fracturing process were not clear. The present experimental procedure by use of initial motions of P-waves is very simple and provides accurate results. By this new technique, a large number of source locations in successive stages in fracturing process have been determined for various rock specimens with different structures."

Onoe (1974), Chairman of the AE research group, Japan, wrote in his general review paper that Mogi's experiment was well-known as the first reliable study in the world, in which the fracture propagation process was tracked by the AE source location method.

Thereafter, a large number of source location studies were carried out using more modern methods (e.g. Lockner and Byerlee, 1977; Nishizawa et al., 1981; Yanagidani et al., 1985; Masuda et al., 1988). Among these studies, the earliest experiment by Mogi (1968) is probably still one of the most accurate source location studies, and it is helpful to understand the process of fracture propagation prior to the main rupture of a rock sample. In this section, this experiment will be explained in detail.

5.2   EXPERIMENTAL PROCEDURE

5.2.a   *Measurement of very high frequency elastic waves*

Fracture of specimens of brittle heterogeneous rocks is preceded by the occurrence of AE events caused by micro-fracturing, as mentioned earlier. These AE events are usually elastic waves of very high frequency, although they also contain a low frequency component. The predominant period of AE events accompanying a micro-fracture 3 mm in length may be roughly estimated 1 μs from (crack length)/(velocity of crack propagation). Amplitudes of elastic waves in AE events vary widely. For determination of the source location of AE events, elastic waves of high frequency need to be recorded, particularly their initial part, and differences in arrival times of elastic waves at different transducers within small rock samples need to be detected.

The synchroscope-moving camera system devised for this experiment makes it possible to observe such elastic waves of very high frequency and to detect a small difference in arrival time, such as 1 μs. The experimental arrangement is schematically shown in Fig. 5.1. The transducers used in the experiment are lead-titanate-zirconate compressional mode disks, 10 mm in diameter and 3 mm thick. To completely observe the initial part of a signal, transducers for triggering synchroscopes and cameras were situated close to an AE source region, and transducers for observation of signals were placed at a distance from the source region. Amplification was achieved by pre- and main-amplifiers of which frequency response was flat from 10 kHz to 1 MHz. Signals traced in two Iwasaki DS-5158 dual-beam synchroscopes were photographed by Borex H16M moving cameras which had been set up for a shot. Just after shooting, a single frame of a film was moved and the cameras were ready for the next shot.

Two traces were recorded in a single frame and a one micro-second arrival time difference between two traces was detectable. For time recording, an analog type watch was photographed on the same film and the time in seconds could be read from

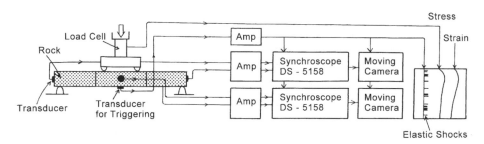

Figure 5.1.   Experimental system for measurement of source location of AE events caused by bending stress (Mogi, 1968).

it. (In 1968, digital type watches were not available, but they were devised in 1973 in Japan.) The applied load and the deformation of rock specimens were measured by strain meters of electric-resistance strain gage type and these were continuously recorded together with pulses showing the occurrence of AE events.

This system has the following features:

1. A micro-second time difference between two traces of the initial part of P-wave is detectable, so that source locations of AE events in a small rock sample can be determined accurately.
2. By suitable setting of transducers for triggering synchroscopes, the wave shape of AE events, including the initial motion of P-wave, can be observed.
3. Multi-trace observations are possible by increasing the number of channels in the same system.

Some examples of observed signals are presented in Fig. 5.2(a) and 5.2(b). The two traces in each single frame in Fig. 5.2(a) are records of the same AE event at different magnification. Two traces in each frame in Fig. 5.2(b) show records of the same AE event from different transducers situated at different locations. From the

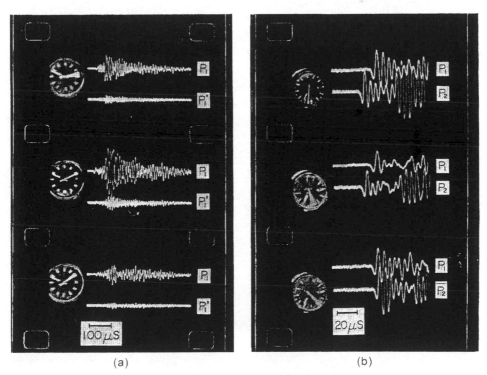

(a)   (b)

Figure 5.2.   (a) (b) Examples of records of AE events in bending experiment of a granite specimen. Arrival times in seconds are indicated by the watch photographed together in each frame. (a) Two wave records from the same transducer by different magnification; (b) Two wave records from different transducers P_1 and P_2. Differences of arrival time of P-wave can be seen clearly.

differences of arrival time of P-waves between the two traces in Fig. 5.2 (b), source locations are determined in millimeters. From these wave records, radiation patterns of elastic waves, focal mechanism, frequency spectrum and other features can also be investigated.

5.2.b   *Determination of source location of AE events*

In the experiment, a bending load was applied to beam-shaped rock specimens. The specimen shape and the loading system are shown in Figs 5.3 and 5.4. By this application of a uniform bending load to the central part of the specimen, the upper and lower sides are stressed in compression and tension, respectively. Since the tensile strength of brittle rocks at atmospheric pressure is about one-tenth of the compressive strength, fractures take place only on the tensile side in the central part, particularly in the surface layer. Thus, if the stress gradient within the beam is ignored, this experiment approximately corresponds to a simple tension experiment, in which the tensile stress on the surface layer is uniform in the central part of the specimen.

5.2.b.1   *One-dimensional case*
At first, one-dimensional source locations of AE events in a direction of tensile stress (OL in Fig. 5.3) were determined. Locations of transducers are shown in Fig. 5.3. P_1 and P_2 are transducers for measurement of signals and P_t are transducers for triggering a synchroscope-camera system. The delay of arrival times of elastic waves of P_2 from P_t makes it possible to observe initial motions of elastic waves from AE sources. Examples of signals observed at P_1 and P_2 in the fracturing process in a granite specimen

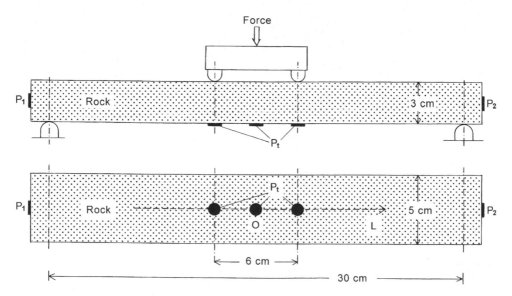

Figure 5.3.   The rock sample and transducers used in studying one-dimensional source locations of AE events in rocks subjected to bending stress. P_1 and P_2: transducers for AE signal observations. P_t: Transducers for triggering a synchroscope and a camera.

are shown in Fig. 5.2(b). Since the velocity of P-waves in this granite is known, source locations of AE events in the OL direction can be determined from the difference in arrival time of P-waves between two traces. As can be seen in this figure, the initial motions of both traces (P_1 and P_2) are always downwards. This result shows that the focal mechanism of these AE events under bending load is similar and these AE events occurred by generation of micro-tensile-type cracks which opened in the OL direction.

5.2.b.2 *Two-dimensional case*

Two dimensional AE source locations were determined by applying the above-mentioned method to two directions. Test specimens used in this case are shown in Fig. 5.4. Measurements in the OL direction are similar to those in the one-dimensional case. Locations in the direction perpendicular to the OL direction were determined from the difference in arrival times between P_3 and P_4. To obtain a delay time at P_3 and P_4, paraffin blocks were stuck to both sides of the rock sample. For the visco-elastic properties of paraffin, the paraffin blocks do not affect the stress distribution, but they

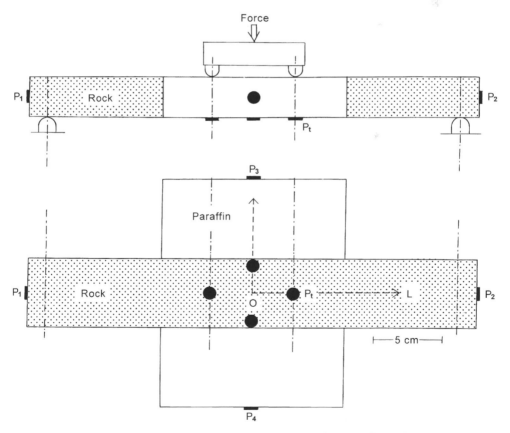

Figure 5.4. The rock sample and transducers used in studying two dimensional source locations of AE events. P_1, P_2, P_3 and P_4: transducers for two-dimensional source locations of AE events. P_t: Transducers for triggering synchroscopes and cameras.

act as good mediums for propagation of elastic waves. Two-dimensional AE source location can be determined based on two differences in arrival times between P_1 and P_2 and between P_3 and P_4 when elastic wave velocity is considered, as mentioned above.

5.3   EXPERIMENTAL RESULTS

As shown in Fig. 5.2, elastic waves originating from AE sources in the stressed rock specimen are quite similar in shape to those of natural earthquakes. In some cases, the S-phase can be clearly identified, but in other cases this phase is not clear. Therefore, the use of the initial motion of P-wave is essential for accurate determination of source location. This situation is very similar to that in the determination of the hypocenters of natural earthquakes. The modern automatic hypocenter determination of natural earthquakes, based on the threshold level of signals, is not always accurate, and so the final reliable results are obtained by revision based on the manual reading of the arrival time difference of P-waves. Since the AE source location in this experiment was determined by manual reading of a clear initial P-phase, this result is quite reliable among many recent source location studies.

In this experiment, three granites, an andesite, a trachyte and three marbles were used. As mentioned in the preceding section, the AE activity increases markedly prior to the main fracture of heterogeneous brittle rocks, and the AE activity greatly depends on the degree of heterogeneity of rock material.

5.3.a   *Granite (heterogeneous silicate rock)*

As heterogeneous brittle rocks, Inada granite, Kitagi granite and Mannari granite were used for AE source location studies. In these rocks, the process of occurrence of AE events before the main rupture is divided into the following three stages: (**A**) – Initial stage in which no appreciable AE events occur; (**B**) – In this stage, AE events begin to occur and their sources are distributed randomly in the specimen; (**C**) – The last stage in which sources of AE events are concentrated in limited regions. Typical examples are explained below.

Figures 5.5–5.7 show source locations in a specimen of Inada granite (I4). The right picture in Fig. 5.5 shows a typical simple pattern of one-dimensional source location in the OL direction as a function of time. Different diameters of solid circles indicate different magnitudes of AE events. The left picture in Fig. 5.5 shows the applied tensile stress as a function of time. At the right side of the right picture, stages **A**, **B**, C_1, C_2 and C_3 are indicated. AE events begin to occur at a certain stress level and increase with the increase of the applied stress. In stage **B**, sources of AE events appear randomly in the direction of tensile stress. In the next stage (**C**), sources begin to concentrate in a certain region and AE events continue to occur in this region up to the main rupture. Activity in other regions decreases gradually with time. And just before the rupture, source locations of AE events are completely limited to the

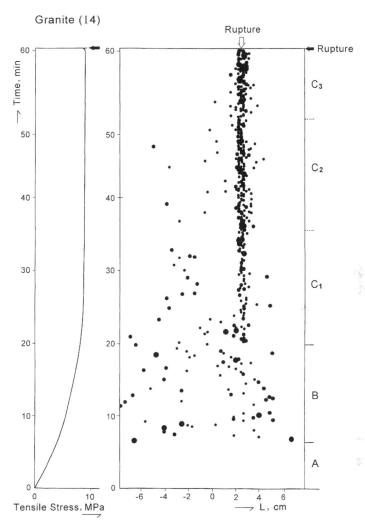

Figure 5.5. Source locations of AE events in the OL direction as functions of time in Inada granite specimen (I4). The applied stress in the tensile side is shown as a function of time in the left figure. Stages **A**, **B** and **C** are shown in the right side.

above-mentioned region where the main rupture will occur. The total frequency distribution of AE sources in the OL direction has a single sharp peak.

Figure 5.6 shows the two-dimensional source locations in successive stages **B**, **C**$_1$, **C**$_2$ and **C**$_3$. In stage **B**, AE sources are distributed at random in the stressed region. At this stage, the location of the impending main rupture cannot be predicted. In stage **C**$_1$, AE events begin to concentrate in a limited region. In stages **C**$_2$ and **C**$_3$, the source region develops in the direction perpendicular to the applied tensile stress up to the main rupture.

The successive development of the active area before the main rupture is represented by the main crack pattern in Fig. 5.7. The highly active areas, the shaded areas in

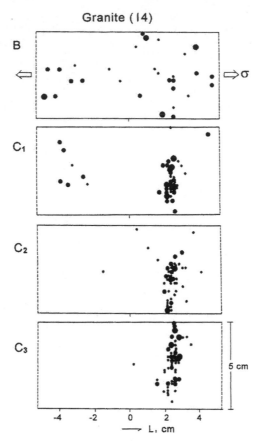

Figure 5.6.   Two-dimensional source locations of AE events in successive stages in Inada granite specimen (I4). Stages **B**, **C**$_1$, **C**$_2$, and **C**$_3$ correspond to those in Fig. 5.5.

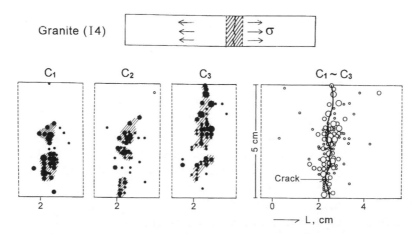

Figure 5.7.   Successive development of the source regions of AE events in the stages **C**$_1$–**C**$_3$ and an observed main crack.

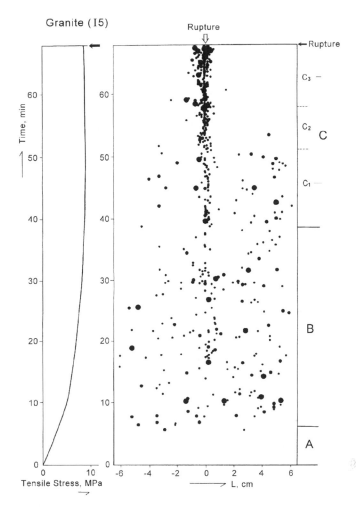

Figure 5.8. Source locations of AE events in the OL direction as functions of time in Inada granite specimen (I5). The applied stress in the tensile side is shown as a function of time in the left figure. Stages **A**, **B** and **C** are shown in the right side.

this figure, successively migrated outward, and an area where the activity had been markedly high in the preceding stage became relatively calm in the subsequent stage. This feature in the fracturing process is also seen in the seismic activity in large-scale fields. In the right picture in Fig. 5.7, the main crack pattern observed after the experiment is compared with AE source locations. The above-mentioned development of the AE source region indicates the successive propagation of the main tensile crack perpendicular to the direction of tensile stress.

Figures 5.8–5.10 show similar AE source locations in an Inada granite specimen (I5). Figure 5.8 shows a typical simple pattern of one-dimensional source location in the OL direction (the direction of tensile stress). AE events begin to occur in stage **B**, and AE sources appear at random in this stage. In the next stage (**C**), AE sources

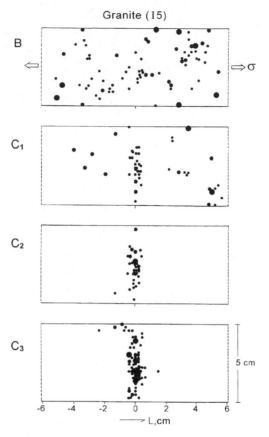

Figure 5.9.   Two-dimensional source locations of AE events in successive stages in Inada granite specimen (I5). Stages **B**, **C₁**, **C₂**, and **C₃** correspond to those in Fig. 5.8.

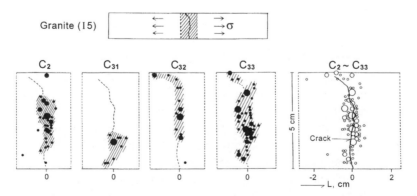

Figure 5.10.   Successive development of the source regions of AE events in the stages C_2, C_{31}, C_{32} and C_{33} and an observed main crack.

concentrate in the limited region where the main rupture will occur. Stages **A**, **B**, **C**$_1$, **C**$_2$ and **C**$_3$ are indicated in the right side of this figure.

The two-dimensional source locations in successive stages are shown in Fig. 5.9. In stage **B**, the AE activity is noticeably higher and source locations are also random and no locations suggesting the future development of the main crack can be found. In the next stage, **C**$_1$, some weak concentration of AE sources can be recognized at the center of the investigated area. The concentration of sources in the limited area becomes evident in the following stage, **C**$_2$, and the AE source area extends to the whole width of the sample in the last stage, **C**$_3$. In Fig. 5.10, the successive development of the AE source region in stages **C**$_2$ and **C**$_3$ is shown with the main crack pattern observed after the experiment. In stage **C**$_{32}$, the AE source region developed upward and this region clearly curved in the left direction near the upper boundary of the specimen. As can be seen in the right picture in Fig. 5.10, the main crack observed after the experiment curves in the left direction near the upper boundary. Thus, the active AE source region agrees very well with the observed crack pattern. This result directly shows the high accuracy of the AE source determination in this experiment.

The above-mentioned two examples are typical simple cases in which the total frequency distribution of AE sources in the direction of tensile stress (OL direction) has a single sharp peak. However, there are also more complex cases.

Figure 5.11 shows a case in which the total frequency distribution of AE sources in the OL direction has two peaks at zones **I** and **II**. The micro-fracturing process is also divided into stages **A**, **B** and **C**. In stage **B**, AE sources are nearly randomly distributed, and at stage **C**$_1$, they concentrate in zones **I** and **II**, and AE source locations at the last stage **C**$_2$, are limited to zone **I**.

In Fig. 5.12, the number of AE events originating from zones **I** and **II** are shown as a function of time, together with that of the total number. The total frequency curve may suggest an apparent stationary occurrence of AE events in stages **B** and **C**$_1$, except for the last stage (**C**$_2$). However, the two lower curves indicate a remarkable change during this fracturing process. In stage **B**, AE activity markedly increased in zone **II**, but it was very low in zone **I**. In stage **C**$_1$, the activity gradually decreased in zone **II**, but it began to increase in zone **I**. At the last stage (**C**$_2$), the activity completely decayed in zone **II**, but increased rapidly in zone **I**. The micro-fracturing activity in zone **I** built up to the main rupture. Thus, adjacent zones closely interact, and the active region migrates to the adjacent region. Migration of seismic activity on a large scale sometimes occurs in natural earthquakes.

Figure 5.13 shows the case of Kitagi granite (K2) in which the pattern of AE source location is more complex. This figure shows AE source locations in the OL direction as a function of time. A total distribution of AE sources along the OL direction has three marked peaks at zones **I**, **II** and **III**, as shown at the top of this figure. In Figure 5.14, the frequency distribution of AE sources along the OL direction in stage (1) (0–20 minutes), stage (2) (20 minutes–40 minutes) and stage (3) (40 minutes–51 minutes) are shown with the total frequency distribution [(1) + (2) + (3)] and the crack pattern observed after the experiment. The above-mentioned three peaks **I**, **II**

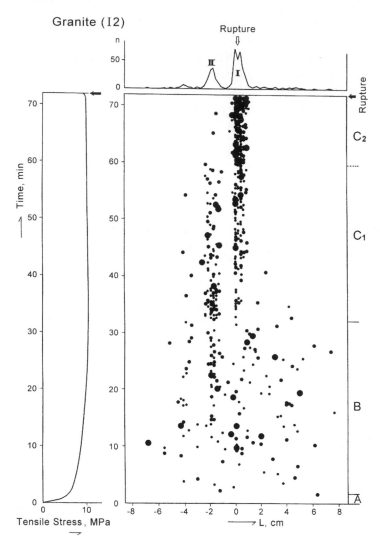

Figure 5.11.   Source locations of AE events in the OL direction as functions of time in Inada granite specimen (I2). The applied stress in the tensile side is shown as a function of time in the left figure. The top figure is the total number (n) of AE events along the OL direction.

and **III** correspond well to the locations of observed cracks. As can be seen in this figure, the most active region was located at zone **I** in the initial stage (1), at zone **II** in the middle stage (2) and at zone **III** in the last stage (3). Thus, the active region systematically migrated with time along the OL direction.

   In Fig. 5.15, the elastic energies (relative values) released as AE events in zones **I**, **II** and **III** are shown as functions of time. The activity in zone **I** abruptly increased in stage (1) and gradually decreased in the following stages. This curve is similar to that of some earthquake swarms. With the decay of the activity in zone **I**, the

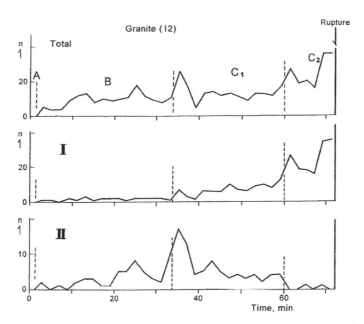

Figure 5.12. Frequency curves of AE events originated from the whole region (top), **I** zone (middle) and **II** zone (bottom). Stages **A**, **B** and **C** correspond to those on Fig. 5.11.

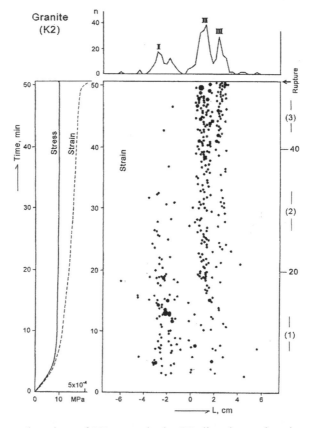

Figure 5.13. Source locations of AE events in the OL direction as functions of time in Kitagi granite specimen (K2). The top figure shows the total number of AE events along the OL direction.

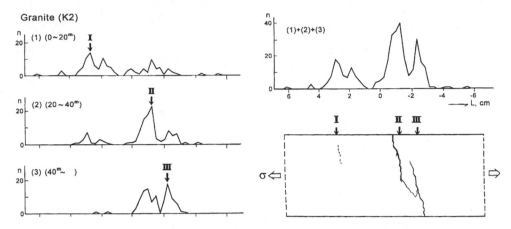

Figure 5.14.   Left figure: distributions of sources of AE events in the OL direction in stages (1), (2), (3); Right figure: the total number of AE events (1) + (2) + (3) and the observed main crack pattern in the rock specimen.

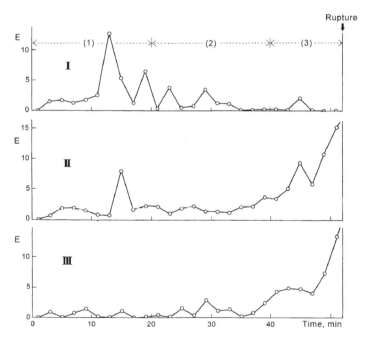

Figure 5.15.   Energies (in relative scale) of AE events originated from **I**, **II** and **III** zones as function of time in Kitagi granite specimen (K2).

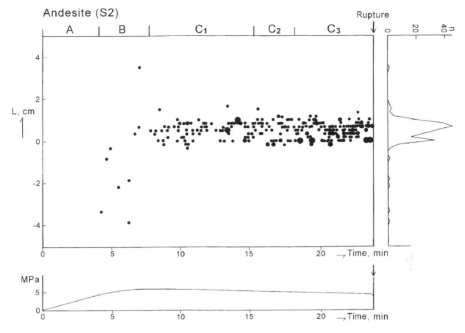

Figure 5.16.   Source locations of AE events in the OL direction as functions of time in Shirochoba andesite specimen (S2). Stage **A**, **B** and **C** are indicated at the top. Right figure: the frequency distribution of AE events along OL line.

activities in zones **II** and **III** increase gradually in stage (2) and markedly in the last stage (3). The curve shapes in zone **I** and zone **III** are contrary to each other, and the curve in zone **II** seems to show an intermediate pattern between the adjacent zones **I** and **III**.

5.3.b   *Andesite (moderately heterogeneous silicate rock)*

The one- and two-dimensional AE source locations of Shirochoba andesite specimen (S2) are shown in Figs 5.16 and 5.17. The degree of heterogeneity of this rock is appreciably lower than that of granites, and so AE activity in the fracturing process is noticeably lower than that of granites. The fracturing process is also divided into three stages: **A**, **B** and **C**. As can be seen in the one-dimensional AE source locations in the OL direction shown in Fig. 5.16, AE activity in stage **B** is very low and AE sources are located in a wide region. In stage **C**, AE activity increases markedly and AE sources are located in a limited region up to the main rupture. The two-dimensional source locations in successive stages are shown in Fig. 5.17. In stage **B**, the AE activity is very low and AE sources are located at random. In stage **C₁**, the AE activity increases and AE sources are located in a limited linear zone perpendicular to the direction of the tensile stress. This active zone developed successively to the upper boundary in the following stages **C₂** and **C₃**, then the main rupture occurred in this region.

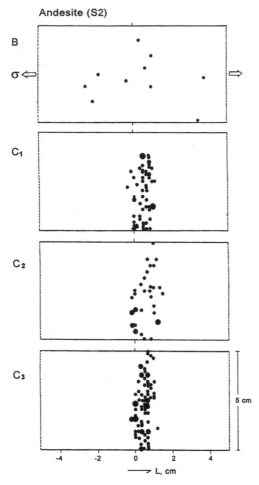

Figure 5.17. Two-dimensional source location of AE events in successive stages in Shirochoba andesite (S2). Stage **B**, C_1, C_2 and C_3 correspond to those in Fig. 5.16.

5.3.c   *Mizuho trachyte (nearly homogeneous silicate rock)*

Mizuho trachyte is a macroscopically nearly homogeneous silicate rock. Three examples of one-dimensional AE source location measurements under the bending load are shown in Fig. 5.18. Only a few AE events were observed before rupture, and the main rupture took place suddenly without any direct relation to the preceding AE events, except for the case of specimen M4. In the case of specimen M4, several AE events began to occur in a limited region and the main rupture occurred in this region. This difference between specimen M4 and the other specimens may be attributed to a slight difference in the degree of heterogeneity (or homogeneity) between these rock specimens. Although the AE activity in Mizuho trachyte is very low, the space-time pattern of AE events of this rock is similar to those of granite and andesite.

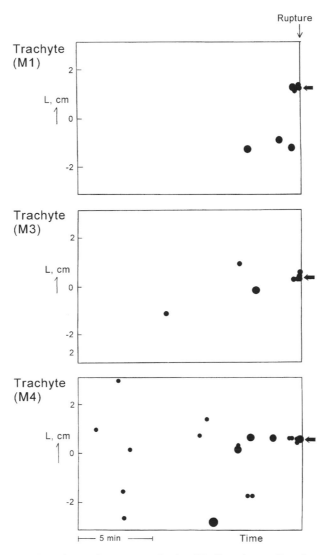

Figure 5.18. Source locations of AE events in the OL direction as functions of time in three Mizuho trachyte specimens M_1, M_3 and M_4. Circles show the magnitude of AE events.

5.3.d *Yamaguchi marble with different grain sizes*

AE source location measurements were carried out on three Yamaguchi marble varieties with different grain sizes: 0.2, 2 and 5 mm. The structure of fine-grained marble is regarded as homogeneous in this investigated scale, and that of coarse-grained marbles is significantly non-uniform. According to previous studies, fracturing behavior depends on the structural heterogeneity; therefore, it was expected that AE source locations would be different for specimens with different grain size.

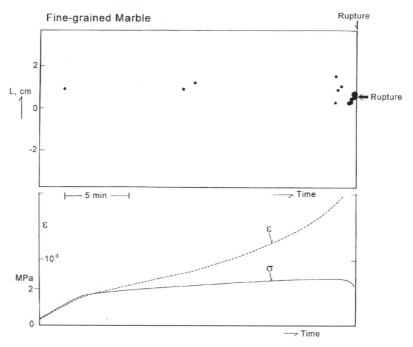

Figure 5.19.   Top figure shows source locations of AE events in the OL direction as functions of time and the bottom figure shows stress (σ) and strain (ε) as functions of time in the fine-grained marble.

Figures 5.19 and 5.20 show typical experimental results for fine-grained marble (YF) and coarse-grained marble (YC), respectively. Each upper picture shows AE source locations in the OL direction as functions of time. Lower picture shows the applied stress and the strain as functions of time. In the fine-grained marble with homogeneous structure, no measurable AE events occur prior to a main rupture. In the coarse-grained marble, however, a number of AE events continued to occur up to the main rupture. In this case, AE sources do not concentrate in a limited region, but they are not distributed at random. As can be seen in Fig. 5.20, the AE activity is not continuous in space and time up to the main rupture. This pattern of AE source locations may be attributed to a relatively large-scale inhomogeneous stress distribution due to the coarse grain size. Fracturing behavior of the medium-grained marble, as suggested by AE source location experiments, is intermediate between the above-mentioned extreme cases.

5.3.e   *Fracture of a semi-infinite body by an inner pressure source*

The source location of AE events generated by the fracturing process of a semi-infinite brittle medium loaded by an inner stress source was studied as a simple model experiment of the generation of a shallow earthquake. The experimental method is schematically explained in Fig. 5.21. The top picture and the bottom picture show

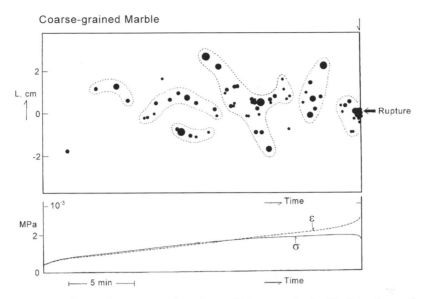

Figure 5.20.   Top figure shows source locations of AE events in the OL direction as functions of time and the bottom figure shows stress (σ) and strain (ε) as functions of time in the coarse-grained marble.

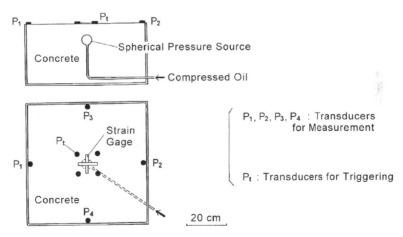

Figure 5.21.   Experiment of source location of AE events in the fracturing process of a semi-infinite medium loaded by an inner stress source: A simple model experiment of earthquake generation.

the vertical section and the horizontal view of the main part of the experimental device. Concrete was used as a brittle medium. A hollow rubber sphere, to which compressed oil is supplied by a pump, is buried at a depth. P_1–P_4 are transducers for AE measurement and P_t is the transducer for triggering the synchroscope-moving camera system, mentioned earlier.

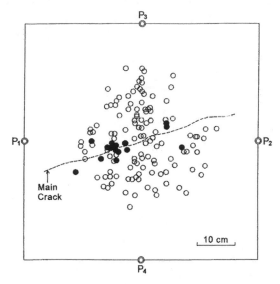

Figure 5.22. Two dimensional AE source locations during the fracturing process of the semi-infinite body. Solid circles indicate AE events prior to the main fracture and open circles indicate AE events after the main fracture.

With a gradual increase of pressure of an inner stress source, AE events begin to occur near the pressure source region. The two-dimensional AE source locations during the fracturing process of this semi-infinite body are shown in Fig. 5.22. The AE events, which occurred prior to the main fracture, are indicated by solid circles. They correspond to foreshocks in natural earthquakes. Following the main fracture, a number of AE events occurred in a wide region. These events are indicated by open circles. A main crack, which was observed after the experiment, is shown by a broken line. It is noteworthy that AE events (solid circles), corresponding to foreshocks, are located in a limited region along the main crack. Earthquake problems will be discussed in detail in Part 3.

To sum up, AE source location study is very helpful in understanding the fracturing process of rocks, particularly the crack propagation process. It should also be noted that the above-mentioned noticeable changes in the space-time distribution of AE sources before the main fracture were observed under nearly constant loading, as shown in the stress-time curves in Fig. 5.5, Fig. 5.8, Fig. 5.11 and Fig. 5.13. This result may help in better understanding of the occurrence of foreshocks in natural earthquakes. An analogous result was obtained by Nishizawa et al. (1981) in a uniaxial compression experiment.

REFERENCES

Harding, S. T. (1970). A critical evolution of microseismic studies. Internal Report RML- IR/ 70-24, Department of Mineral Engineering, the Penn. State Univ.

Hardy, H. R. (1972). Application of acoustic emission techniques to rock mechanics research. In: Acoustic Emission. Am. Soc. Inst. Mat., St. Tech. Publ. 505, 41–83.

Lockner, D. and J. D. Byerlee. (1977). Acoustic emission and fault formation in rocks. In: Proc. First Conf. on Acoustic/Microseismic Activity in Geologic Structure and Materials. Penn. State Univ., 1975. Hardy, H. R. Leighton, F. W.(eds.). Clausthal: Trans. Tech. Publ., 99–107.

Masuda, K., H. Mizutani, I. Yamada and Y. Imanishi. (1988). Effect of water on time-dependent behavior of granite. J. Phys. Earth, **36**, 291–313.

Mogi, K. (1968). Source locations of elastic shocks in the fracturing process in rocks (1). Bull. Earthquake Res. Inst., Tokyo Univ., **46**, 1103–1125.

Nishizawa, O., K. Kusunose and K. Onai. (1981). A study of space-time distribution of AE hypocenters in a rock sample under uniaxial compression. Bull. Geol. Surv. Japan, **32**(9), 473–486.

Onoe, M. (1974). Fundamentals and Application of Acoustic Emission, Chapter 1 Review, 1-24, Japanese Soc. Non-Destructive Inspection, Tokyo. (in Japanese)

Scholz, C. H. (1968). Experimental study of the fracturing process in brittle rock. J. Geophys. Res., **73**, 1447–1454.

Yanagidani, T., S. Ehara, O. Nishizawa, K. Kusunose and M. Terada. (1985). Localization of dilatancy in Oshima granite under uniaxial stress. J. Geophys. Res., **90**, 6840–6858.

CHAPTER 6

Magnitude–frequency relation of AE events

6.1 INTRODUCTION

In natural earthquakes, it is well known that generally small earthquakes occur frequently, but large earthquakes occur rarely. Gutenberg and Richter (1944) found the following statistical formula about the relation between the magnitude of earthquakes and their frequency:

$$\log N = c + b(8 - M) \tag{6.1}$$

where N is the number of earthquakes and M is the magnitude, introduced by Richter (1935) based on the records made by seismographs.

Ishimoto and Iida (1939) also found the following statistical formula between the magnitude and its frequency of earthquakes:

$$n(a) = Ka^{-m} \tag{6.2}$$

where a is the maximum trace amplitude of the seismogram recorded at an observatory, $n(a)\ da$ the number of earthquakes having a maximum trace amplitude a to $a + da$, and K and m are both constants. Asada, Suzuki and Tomoda (1951) revealed that Eq. (6.2) was equivalent to Eq. (6.1) under a simple assumption and the constant b had the following relation to the exponent m,

$$b = m - 1 \tag{6.3}$$

And it was noticed that the exponent m or b value is roughly constant in many regions, except for some cases.

The first laboratory measurement on the magnitude–frequency relation of AE events in rock fracture was carried out by Vinogradov (1959). Then, Mogi (1962a, b) conducted a series of experiments on this problem and discussed the physical meaning of the magnitude–frequency relation of AE events. Since natural earthquakes are regarded as large scale AE events, those results of laboratory experiments may be useful for the physical interpretation of the observed magnitude–frequency relation of earthquakes.

6.2 EXPERIMENTAL PROCEDURE AND SPECIMEN MATERIALS

The present author studied the magnitude–frequency relation of AE events of rocks and rock-like brittle various materials using the Ishimoto–Iida's method (Mogi, 1962a, b; 1980, 1981).

Preliminary experiments suggested that the magnitude–frequency relation of AE events was strongly influenced by the homogeneity or inhomogeneity of stress distribution in the medium. Therefore, experiments were conducted for examination of the following two effects:

(i) Effect of structure of the medium: specimens of homogeneous or inhomogeneous and regular or irregular structures were used.
(ii) Effect of applied stress: the stress was applied uniformly in one case, but was applied non-uniformly in another case.

Thus, AE events in the fracturing processes of brittle materials with different structures caused by different types of applied stresses were measured. Figure 6.1 shows a

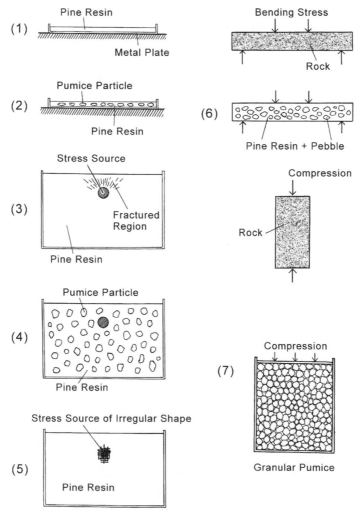

Figure 6.1.   Schematic views of the states of materials and the applied stress in the experiments discussed in this chapter.

schematic view of various cases of the state of materials and the applied stresses in the experiment. In the early works (Mogi, 1962a,b), relatively low frequency components (100–1000 Hz) of acoustic waves (AE) were measured, and then both low and high frequency components were measured (Mogi, 1980, 1981).

As mentioned above, Fig. 6.1 shows schematically the state of specimens which were used in the early experiments (Mogi, 1962a,b) which are explained below. In these experiments, pine resin was used frequently with brittle rocks, because pine resin is a typical brittle material at room temperature and easy to handle.

(1) *A sheet of homogeneous material (pine resin) stressed by a uniform thermal stress.* The pine resin was solidified in a circular metal plate which is shown in Fig. 6.1 (1). The diameter of the circular sheet of pine resin is 20 cm and its depth is 1–3 mm. Initially the metal plate was heated and the pine resin melted in the metal plate, and then the system was cooled slowly. Since the coefficient of thermal expansion of pine resin (in the solid state) is quite different from that of the metal sheet, the thermal stress (tension) is applied to the pine resin at the contact surface with the metal plate in the cooling process. When the cooling of the pine resin is not quite uniform, cracks begin to appear in a limited region, and then the cracks develop gradually outwards and AE events occur successively. In this case, log a–log n relation of AE events is approximately linear and the exponent m is relatively small (0.3–1.2). On the other hand, when the cooling of the medium is quite uniform, a very regular crack pattern (see Fig. 6.9) appears suddenly over the whole area, and many small AE events caused by small local cracks follow. In this case, the log a – log n relation deviated from a straight line concavely downwards. This case will be discussed in a later section.

(2) *A sheet of heterogeneous material stressed by uniform thermal stress.* The mixture of pine resin and volcanic pumice particles, which was solidified in a circular metal plate, was presented as a heterogeneous material, as shown in Fig. 6.1 (2). The

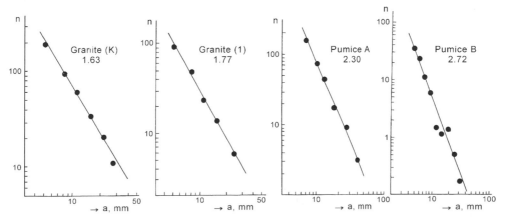

Figure 6.2.   Relation between the maximum trace amplitude a of AE events and its number $n(a)$ in granites and granular volcanic pumices.

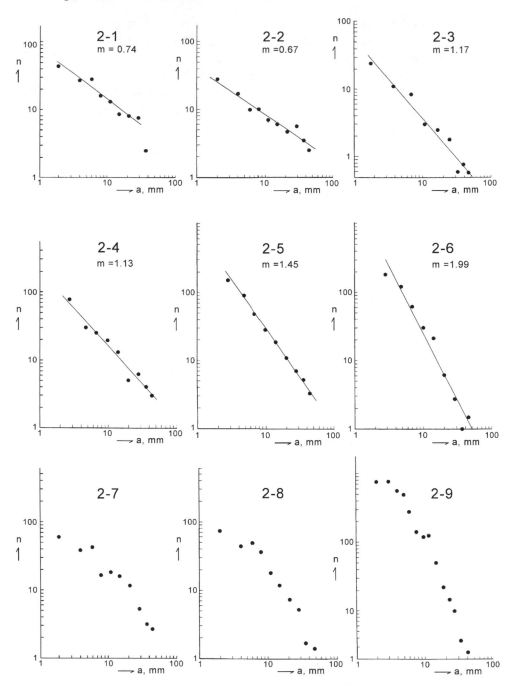

Figure 6.3.   Relation between the maximum trace amplitude *a* of AE events and its number *n(a)* in Experiment (2). 2-1: $r = 0.17$, 2-2: $r = 0.33$, 2-3: $r = 0.33$, 2-4: $r = 0.33$, 2-5: $r = 0.66$, 2-6: $r = 1.0$, 2-7: $r = 0.33$, 2-8: $r = 0.66$, 2-9: $r = 1.0$.

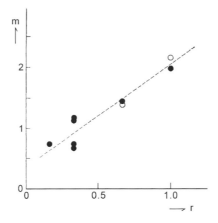

Figure 6.4.   Relation between the *m* value and the mixture ratio *r*.

degree of heterogeneity varies with a mixture ratio of pumice particles to the pine resin medium. The mixture ratio *r* is defined by

$$r = V_P/V \tag{6.4}$$

where V is the total volume of the mixture and V_P is the apparent volume of granular pumice. In the cooling process, AE events accompanying local fractures increase gradually in number. The magnitude–frequency relations of AE events in this case are shown in Fig. 6.3. They are expressed by a straight line, and the exponent *m* increases clearly with the mixture ratio *r* (Fig. 6.4).

(3) *Homogeneous brittle medium stressed by an inner stress source with a round shape.* This experimental procedure is shown schematically in Fig. 6.1 (3). When the inner stress increases gradually, a large fracture occurs suddenly in the stressed region and then small fractures occur successively. During this fracture process, many AE events occur. The log *a* – log *n* relations are frequently linear and some trend concavely downwards. The exponent *m* in this case is in the range 0.9 to 1.7.

(4) *Heterogeneous brittle medium stressed by an inner stress source with a round shape.* The heterogeneous structure is obtained by mixing pumice particles with the pine resin medium, and the mixture ratio *r* is 0.8–1 in this case. The number of AE events gradually increases with increase of the applied stress. In this case, log *a* – log *n* relations are mainly linear.

(5) *Homogeneous medium stressed by an inner stress source with an angular shape.* The shape of the stress source is very angular and many stress concentrated points appear around the source (Fig. 6.1 (5)). Therefore, the stress distribution in the medium is very complex. In this case, AE events increase gradually with time under the increasing stress. log *a* – log *n* relations are also nearly linear and the exponent *m* is somewhat large (1.7–2.0). This result indicates that the spatial variation in the applied stress distribution has also an important influence on the magnitude–frequency relation of AE events.

(6) *Heterogeneous material under uniform stress.* $\log a - \log n$ relations of heterogeneous rock specimens (e.g. granite, andesite) under uniform (or nearly uniform) compressive or bending stress are nearly linear and the exponent m is 1.6–2.0.

(7) *Granular materials under compression.* Under compression, many stress concentrations take place in media of granular structures. Volcanic pumice is porous and complex in structure. Therefore, when granular pumice is compressed, as shown in Fig. 6.1 (7), the stress distribution fluctuates extremely. In this case, $\log a - \log n$ relations are linear and the exponent m is very large (2.3–2.7). Figure 6.2 shows typical examples of $\log a - \log n$ relations of granites and granular pumice specimens. These results show that the magnitude–frequency relation of AE events in both cases are well expressed by the Ishimoto–Iida equation (Eq. 6.2), and the exponent m is significantly different between the cases of granite and pumice.

6.3   THE m VALUE IN THE ISHIMOTO-IIDA EQUATION

As mentioned above, the magnitude–frequency relations of AE events are frequently expressed by Eq. (6.2), and the exponent m varies over a wide range. Figure 6.3 shows relations between the maximum trace amplitude (a) of AE events and its number $n(a)\,da$ in Experiment (2). In this case, the mixture of pine resin and pumice particles which was solidified in a circular metal plate was used, as mentioned above. The degree of heterogeneity varies with a mixture ratio r of pumice particles to the pine resin medium. In this case, it is important that the diameter of granular pumice particles is larger than the thickness of the pine resin medium (see Fig. 6.1 (2)).

Figure 6.4 shows the relation between the m value and the mixture ratio r. Solid circles show the case in which the $\log a - \log n$ relation is quite linear and so the reliability of the m value is high. Open circles show the case in which the linearity is not so high. Anyway, this figure shows clearly that the m value increases with the degree of the structural heterogeneity (r).

This conclusion is also supported by the following experimental results. When many cracks take place successively in the medium in Experiment (2), the density of cracks in the medium increases with time. Figure 6.5 shows the magnitude–frequency relations of AE events in successive stages (1) 0–30 sec, (2) 30–60 sec, and (3) 60–120 sec using the media with three different mixture ratios ($r = 0.33$, 0.66 and 1.0) in Experiment (2). The numerals in each curve show the m values. In Fig. 6.6, it is shown that the exponent m values in successive stages increase markedly with time. These experimental results may be explained by the following simple model: Since the development of cracks is stopped by barriers, the m value increases with the increase of structural heterogeneity, which corresponds to the density in the spatial distribution of mechanical barriers, that is, the inhomogeneity of the stress distribution.

The m values in various mechanical states are summarized in Table 6.1 and Fig. 6.7. They include results of the author's other experiments and results obtained by

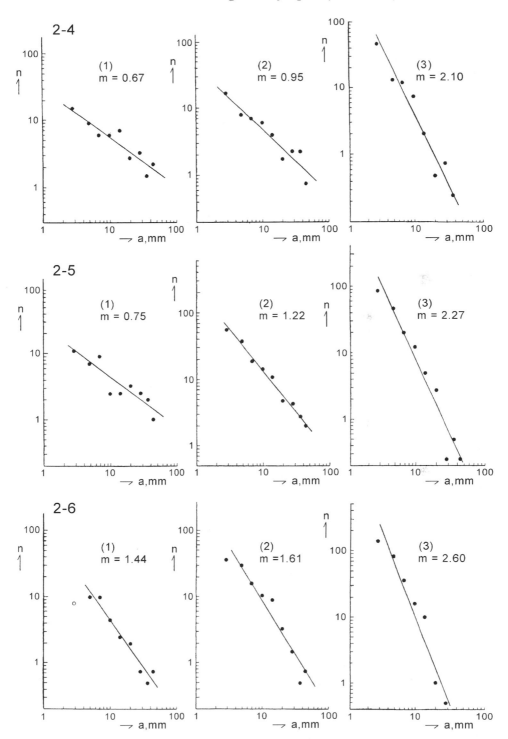

Figure 6.5. Magnitude–frequency relation of AE events in successive states in Experiment (2). Stage (1): 0 sec–30 sec, Stage (2): 30 sec–60 sec, Stage (3): 60 sec–120 sec. 2-4: $r = 0.33$, 2-5: $r = 0.66$, 2-6: $r = 1.0$.

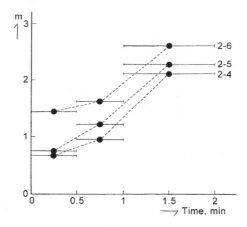

Figure 6.6. Changes in the *m* value for successive states (Experiment (2)). 2-4: $r = 0.33$, 2-5: $r = 0.66$, 2-6: $r = 1.0$.

Table 6.1. Experimental results on the magnitude–frequency relation.

Number of experiment	Material (structure)	Stress application	Applied stress states	Type of log n–log a relation	m^*	Remark
(1)-1	pine resin	uniform	nearly	I	0.66	$d^{**} = 0.4$ mm
-2	(homogeneous)	thermal	uniform	I	0.86	0.8
-3		stress		I	1.15	1
-4				II	(0.34)	2.8
-5				II	(0.40)	1.3
-6			completely	III		0.15
-7			uniform	III		0.15
-8				II		2.0
(2)-1	mixture of pine	uniform	uniform	I	0.74	$r^{***} = 0.17$
-2	resin and pumice	thermal		I	0.67	0.33
-3	particles	stress		I	1.17	0.33
-4	(heterogeneous)			I	1.13	0.33
-5				I	1.45	0.66
-6				I	1.99	1
-7				II	(0.8)	0.33
-8				II	(1.4)	0.66
-9				III	(2.2)	1.0
(3)-1	pine resin	stress by an	nearly	I	0.93	
-2	(fractured region:	inner source	uniform	I	1.08	
-3	heterogeneous)	of round		I	1.12	
-4		shape		I	1.36	
-5				I	1.54	
-6				I	1.60	
-7				I	1.55	
-8				I	1.78	
-9				II		
-10				I	1.66	

(Continued)

Table 6.1. (*Continued*)

Number of experiment	Material (structure)	Stress application	Applied stress states	Type of $\log n$–$\log a$ relation	m^*	Remark
(4)-1	mixture of pine	stress by an	nearly	I	1.01	$l^{****} = 5$–8 mm
-2	resin and pumice	inner source	uniform	I	1.13	1–2
-3	particles	of round		I	1.13	5–8
-4	(heterogeneous)	shape		I	1.20	1–2
-5				I	1.39	5–8
-6				I	1.52	5–8
-7				I	1.57	2–3
-8				I	1.84	5–8
-9				I	1.35	5–8
-10				I	1.12	5–8
-11				I	1.20	3–5
-12				I	1.31	3–5
-13				I	1.34	3–5
-14				I	1.40	2–3
-15				I	1.45	2–3
-16				I	1.52	1–2
-17				II		5–8
(5)-1	pine resin	stress by an	not	I	1.98	
-2	(homogeneous)	inner source	uniform	I	1.77	
-3		of irregular shape		I	1.80	
(6)-1	mixture of pine	bending	nearly	II	(2.14)	$l = 3$–5 mm
-2	resin and pebbles		uniform	II	(2.03)	1–2
-3	(heterogeneous)			II	(1.85)	5–8
-4				II	(1.93)	1–2
-5				II	(1.98)	5–8
-6				I	1.84	2–3
G(M)	granite	bending	nearly	I	1.61	$l = 0.1$–5 mm
G(I)	granite		uniform	I	1.74	
G(K)	granite			I	1.68	
A(M)	andesite (heterogeneous)			I	2.03	
	granite	compression	uniform	I	1.65 ⎫	calculated
	diabase			I	1.65 ⎬	from
	coal			I	1.90 ⎭	Vinogradov's data
(7)-1′	coal (granular)	compression	uniform	I	1.75	
-1	(very heterogeneous)			I	2.47	$l = 3$–4 mm
-2	pumice (granular)			I	2.30	3–4
	pumice B			I	2.38	
	pumice B			I	2.20	B: increasing stress
	pumice B			I	2.44	A: constant
	pumice B			I	2.33	stress
	pumice A			I	2.65	

*) m: the exponent in the Ishimoto-Iida equation. When the fracturing occurs successively, the m value in a period from 0 to 2 min. is taken here. **) d: thickness of the pine resin medium. ***) r: mixing ratio. ****): grain size of pumice particles or pebbles.

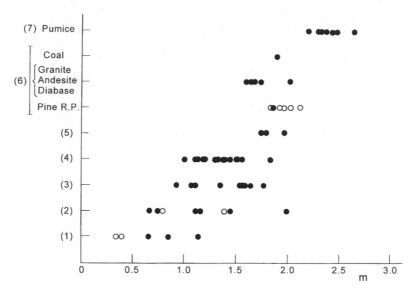

Figure 6.7.   The exponent *m* values in various cases. The numbers in the left side correspond to those in Fig. 6.1.

Vinogradov (1959) for coal, granite and diabase samples. It can be seen that the exponent *m* varies systematically with the structural state of the materials, that is, the *m* value is large (2.1–2.7) for clastic or very porous materials, moderate (1.0–2.0) for compact heterogeneous ones, and very small (0.3–1.0) for nearly homogeneous materials under uniform stress states.

6.4   TYPES OF MAGNITUDE–FREQUENCY RELATIONS AND THE STRUCTURE OF THE MEDIUM

As mentioned above, Eq. (6.2) is not always satisfactory. In general, the magnitude–frequency relation for AE events may be classified into the following types (Fig. 6.8):

   (I) The $\log a$–$\log n$ relation is expressed by a single straight line.
  (II) The $\log a$–$\log n$ relation is expressed by two straight lines.
 (III) The $\log a$–$\log n$ relation is expressed by three straight lines or the relation is expressed by a curved line.

   As mentioned above, the magnitude–frequency relation of AE events is, in many cases, Type I which satisfies Eq. (6.2). Type II is discussed below. In this case, the exponent *m* in the $\log a$–$\log n$ relation curve increases step-wise at a maximum trace amplitude a_d. According to the present author's model (Mogi, 1962a), the exponent *m* is proportional to the probability of which the development of fracture is stopped by obstacles in the medium. In the medium with regular structures shown in Fig. 6.9, this probability is not constant over the whole range of maximum amplitudes of the

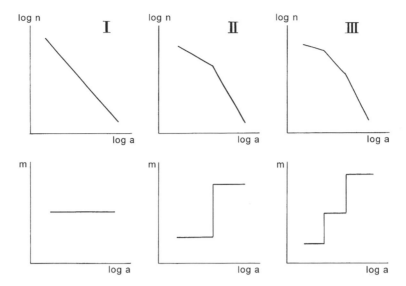

Figure 6.8.   Three typical log *n*–log *a* relations of AE events. The lower figures show three different types of the relation between the gradient d (log *n*)/d (log *a*) ($= m$) and log *a*.

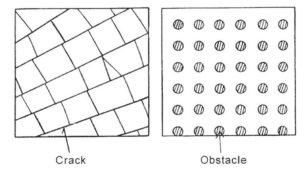

Crack   Obstacle

Figure 6.9.   Examples of mediums with regular structures. Left: regular crack pattern of a pine resin sheet (a case of Fig. 6. 1 (1)), right: pine resin sheet in which obstacles are regularly distributed (a case of Fig. 6.1 (2)).

acoustic waves, and increases suddenly at a certain amplitude (a_d), which corresponds to the dimension of the regular structure.

Figure 6.10 (1), (2) and (3) show log *a*–log *n* relations of the following three successive stages of occurrence of AE events. Stage (1) is 0 sec to 30 sec, Stage (2) 30 sec to 60 sec, and Stage (3) 60 sec to 120 sec. During the cooling process, a very regular pattern of cracks occurred suddenly by thermal stress, over the whole area of a sheet of pine resin in a circular metal plate, and then many minor fractures continued to occur. In Stage (1), the main structure was very regular and the dimension of the unit structure was large, but in the later stages, the regularity of the structure became indistinct and the dimension of the unit structure decreased due to successive fracturing. The value of a_d decreases significantly with the elapsed time and the

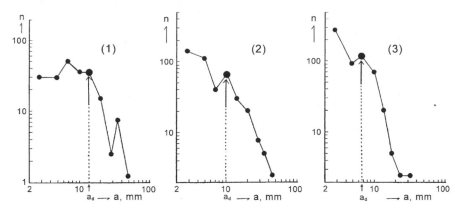

Figure 6.10.   The magnitude–frequency relations of AE events in successive stages for the medium (pine resin sheet) with an initially regular pattern which is the case of the left figure in Fig. 6.9. (1): 0 sec–30 sec, (2): 30 sec–60 sec, (3): 60 sec–120 sec.

discontinuity of the magnitude–frequency relation curve becomes indistinct as Type III. In the later stage, the magnitude–frequency relation becomes linear.

From the above-mentioned experimental results, it is deduced that a linear $\log a - \log n$ relation is obtained when the mechanical structure is very irregular and complex, that is, fractal. This is the reason why the Ishimoto-Iida statistical relation Eq. (6.2) can be applied widely to the magnitude–frequency relations of AE events in fracturing of various materials, including the earth's crust, because the mechanical structure is heterogeneous at various scales. As mentioned above, since the discontinuous point of the $\log a - \log n$ relation curve is related to the dimension of the regular structure, the magnitude–frequency relation of earthquakes seems to give some information about the mechanical structures of a seismic region or stress states in the earth's crust.

6.5   EFFECTS OF MEASUREMENTS BY DIFFERENT FREQUENCIES AND DIFFERENT DYNAMIC RANGES OF ACOUSTIC WAVES

In the laboratory experiments mentioned above, the maximum amplitudes of acoustic waves (AE) and the number of AE events were measured in the low frequency range (100 Hz–2 kHz) and in the limited dynamic range of amplitude measurements (e.g. Mogi, 1962 a, b). Figure 6.12 shows schematically the two different $\log a - \log n$ relation curves. The one in the left figure is exactly linear, but the one in the right figure is curved. This difference is quite important in the discussion on the magnitude–frequency relation problem related to earthquakes (AE events in the earth). However, it seems to be difficult to recognize the difference between these two cases by measurements of acoustic waves over a small dynamic range, as indicated in this figure. Therefore, the magnitude–frequency relation of AE events in the laboratory was measured over a large dynamic range (Mogi, 1980). Scholz (1968) criticized the present author's papers on the magnitude–frequency relation of AE events by measurements

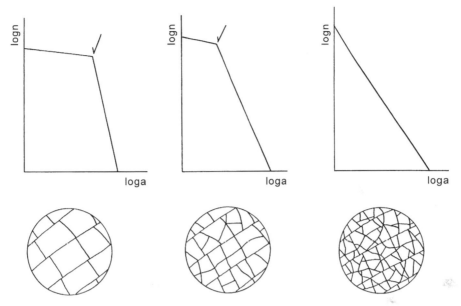

Figure 6.11. The crack patterns (bottom figures) and the magnitude–frequency relations (top figures) in the three successive stages shown in Fig. 6.10.

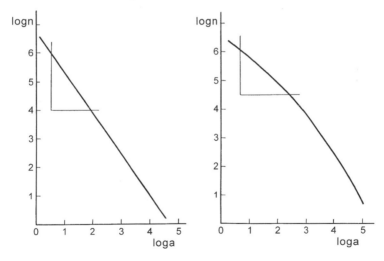

Figure 6.12. Importance of large dynamic range in the measurement of the magnitude–frequency relation. The left figure shows the linear relation, but the relation curve in the right is not linear and concave downward.

in the *low frequency range* of acoustic waves. He wrote that "they (Mogi, 1962, and others) were limited to the *audio frequency range*. Because the frequency content of microfracturing events is primarily much higher than this, his conclusions are open to question." According to his paper, he used a system which had a flat frequency response from 100 Hz to 1 MHz. The audio frequency range is about 20 Hz–20 kHz.

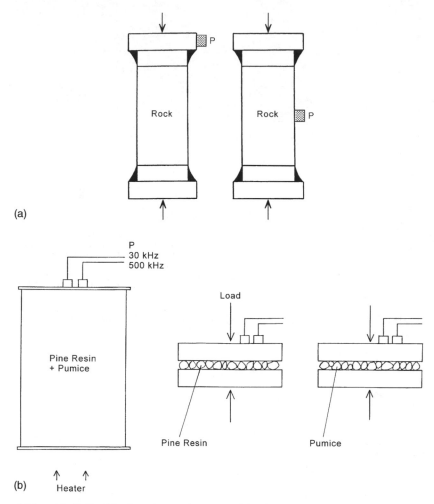

Figure 6.13.   (a) Cylindrical rock specimen used in the measurement of AE events under uniaxial compression. The stress concentration at the end part of the specimen is markedly reduced by the steel-epoxy fillet, as mentioned in Chapter 1. P: pick up, (b) Granular pine resin, volcanic pumice particles and the cylindrical pine resin specimen which contains pumice particles. These specimens are very complex in structure. Two pick ups for different frequency ranges are used.

However, in our early experiments, only larger AE events were measured because of lower sensitivity of the apparatus. Therefore, the effect of measurements in the low frequency range suggested by Scholz, do not cause problems for our conclusion, as explained in the followings. The present author conducted careful experiments on the $\log a - \log n$ relations in two different frequency ranges and examined this problem (Mogi, 1980, 1981).

Figure 6.13 (a) and (b) show the specimens used and the various methods of stress application in the experiment. AE signals were measured in two different period ranges which have characteristic periods through *low* (30 kHz) and *high* (500 kHz) band-pass

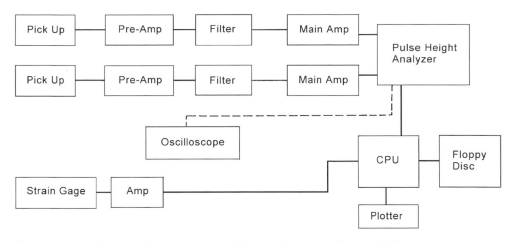

Figure 6.14.   Diagram of measurements of the maximum amplitude of AE event, their number and the strain curves of the compressed medium.

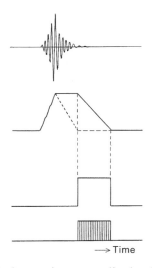

Figure 6.15.   Method by which the maximum amplitude of acoustic wave single of an AE event is converted to an electric signal.

filters. Figure 6.14 shows the diagram of measurements of the maximum amplitudes of acoustic waves, the number of AE events and the strain (deformation) curve of the compressed medium. Figure 6.15 shows the method by which the maximum amplitude of the acoustic wave signal of an AE event is converted to an electric signal. The maximum amplitude corresponds to the duration time of the electric signal. Figure 6.16 shows the frequency characteristics of four different band-pass filters. In the following, the lowest frequency band-pass filter (30 kHz) and the highest frequency band-pass filter (500 kHz) were used.

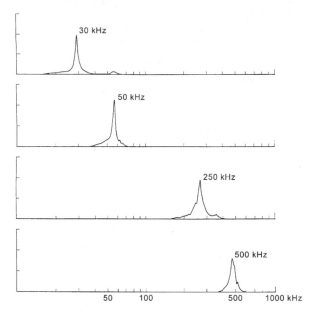

Figure 6.16.   Frequency characteristics of four different band-pass filters used in the experiment, including frequency characteristics of pick ups.

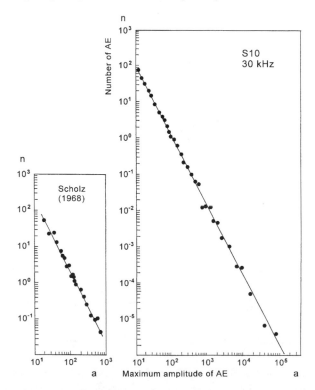

Figure 6.17.   Examination of the linearity of $\log a - \log n$ relation of AE events in fracturing of a heterogeneous material (pine resin + pumice particles). The maximum amplitude of acoustic wave of AE was measured by the large dynamic range system (Mogi, 1980). The left figure from Scholz (1968).

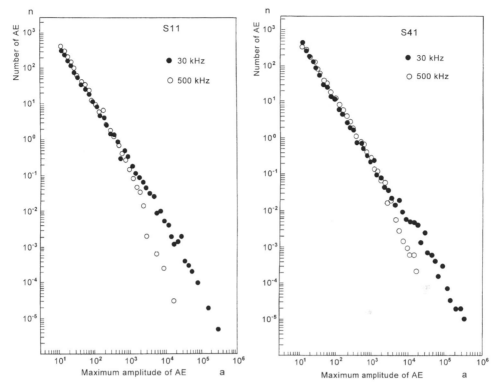

Figure 6.18. $\log a - \log n$ relation of AE events for different frequency ranges. Solid circle: low frequency (30 kHz), open circle: high frequency (500 kHz). Specimen: S11 (pine resin + pumice particles).

Figure 6.19. $\log a - \log n$ relation of AE events for different frequency ranges. Solid circle: low frequency (30 kHz), open circle: high frequency (500 kHz). Specimen: S11 (pine resin + pumice particles). Specimen: S41

Figure 6.17 (right) shows the magnitude (*a*)–frequency (*n*) relation of AE events under uniaxial compression of the cylindrical specimen (S10: mixture of pine resin and pumice particles) (Mogi, 1980). The dynamic range in this measurement of the amplitudes of acoustic waves is exceptionally large for laboratory work. The left figure shows a typical experimental results, reported by Scholz (1968). The $\log a - \log n$ relation in the right figure, obtained using the low frequency band-pass filter (30 kHz) is exactly linear, and the applicability of the Ishimoto-Iida's statistical equation Eq. (6.2) was reconfirmed on the laboratory scale.

The next subject is how the relation between the maximum amplitudes of acoustic waves (AE) and the number of AE events is different for different frequency ranges of acoustic waves. Figures 6.18 and 6.19 show $\log a - \log n$ relations of AE events in the two different frequency ranges. Solid circles indicate the number of AE events counted through the low-frequency band-pass filter (30 kHz) and open circles indicate those through the high frequency band-pass filter (500 kHz). S11 and S41 are the cylindrical pine resin specimens which contain dense pumice particles. Therefore, this material

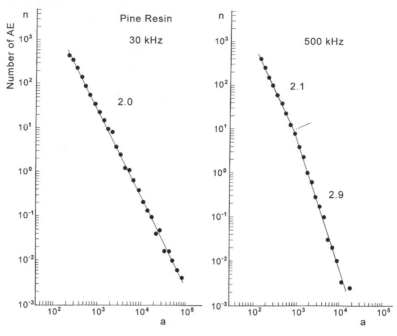

Figure 6.20.   log *a*–log *n* relation of AE events for different frequency ranges (30 kHz and 500 kHz). Specimen: granular pine resin particles.

is very heterogeneous. In these two figures, it is noted that the log *a*–log *n* relation in the low frequency range, indicated by solid circles, is quite linear, but in the high frequency range, indicated by open circles, it is not linear. In the smaller amplitude range, the relation for high frequency (500 kHz) of acoustic waves agrees with that for low frequency (30 kHz), but the slope of the relation for high frequency (500 kHz) is steeper than that for 30 kHz, in the larger amplitude range.

Figure 6.20 shows the log *a*–log *n* relations of AE events for different ranges of 30 kHz (left figure) and 500 kHz (right figure) for the case in which the granular pine resin particles are compressed (see Fig. 6.13 (b)). The *m*-values are larger than in the other cases, however, the log *a*–log *n* relation is quite linear in the low frequency (30 kHz) range, but it is not linear in the high frequency (500 kHz) range. Figure 6.21 shows a summary of the above-mentioned experimental results. These results show that the above-mentioned criticism by Scholz (1968) about the our early AE study (including the magnitude–frequency relation) in the low frequency range is clearly unsuitable.

Figure 6.22 (left) shows the magnitude (*a*) and frequency (*n*) relation in the compressed granular pumice specimen under dry and wet conditions. Solid and open circles show the case of dry and wet condition, respectively. The right figure shows the results for the accumulated number of AE events. According to the experimental results, there is no effect of water for the magnitude–frequency relation of AE events in the volcanic pumice specimen.

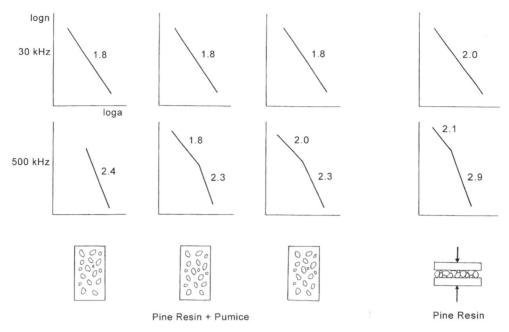

Figure 6.21.   Simplified experimental results of the log *a*–log *n* relation of AE events in the two different frequency ranges.

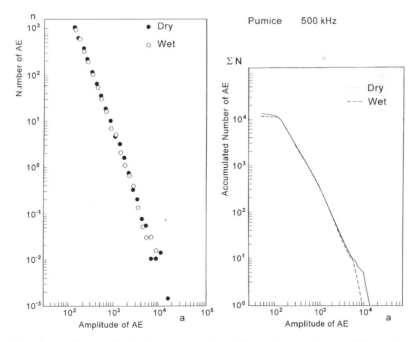

Figure 6.22.   log *a*–log *n* relation in compression of granular pumice particles under dry and wet conditions.

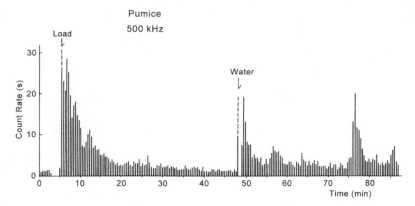

Figure 6.23. Temporal variation of AE activity of granular pumice under constant load conditions. The AE activity changes markedly when water is poured into the specimen.

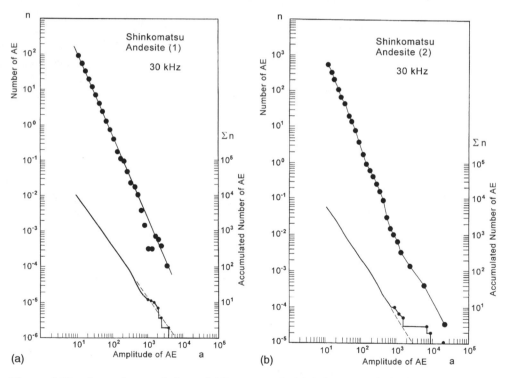

Figure 6.24. $\log a - \log n$ relation of AE events of Shinkomatsu andesite specimens under the uniaxial compression; (a) Specimen (1), (b) Specimen (2). Thick solid line: accumulated number of AE events.

Figure 6.23 indicates that the AE activity of the pumice specimen under a constant load suddenly increases by adding water and then the temporal variation of the AE activity shows significant fluctuation. Thus, the effect of water on the AE activity in rocks is significant.

Shinkomatsu Andesite (1)

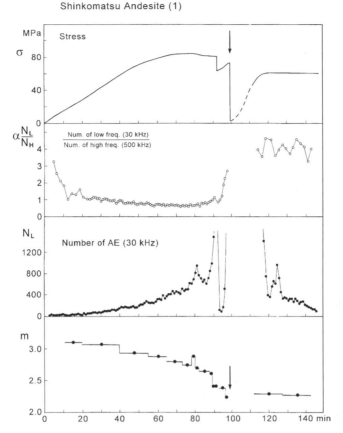

Figure 6.25. Temporal variations of uniaxial compressive stress, ratio of number of AE events of 30 kHz and 500 kHz, number per min. of AE events (30 kHz) per min., and the *m* value, from the top to the bottom. A vertical thick arrow indicates the main fracture. Specimen: Shinkomatsu andesite (1).

Figure 6.24 (a) and (b) show two examples of the magnitude–frequency relation for cylindrical rock specimens under uniform uniaxial compression. The rock used was a porous Shinkomatsu andesite, and the maximum amplitudes of acoustic waves (*a*) was measured through the low band-pass filter (30 kHz). Thick solid lines indicate the accumulated number of AE events. In this rock, the $\log a$–$\log n$ relation is linear for a large dynamic range of amplitude measurements and the *m* value is significantly large.

Figures 6.25 and 6.26 show results of the uniaxial compression experiments under a constant strain rate. The temporal change of σ (applied compressive stress), $\alpha = N_L/N_H$ (ratio of the number of AE events in the low frequency range (30 kHz) and the number of AE events in the high frequency range (500 kHz), N_L (the number of AE events in the low frequency range (30 kHz)), and the *m* value, are shown from top to bottom. In those figures, it is noted that the number of AE events counted

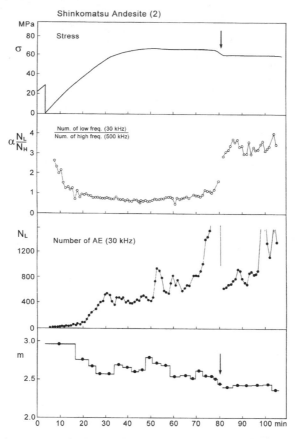

Figure 6.26.   Shinkomatsu andesite specimen (2). The same as Fig. 6.25.

through the band-pass filter of 30 kHz is nearly equal to that of 500 kHz, during the main process until the occurrence of a principal fracture. In this experiment, the compressive stress initially increased at a constant rate and then continued at nearly a constant value. Under the constant compressive stress condition, the exponent m-value of the Ishimoto-Iida's formula gradually decreased until the main rupture. This result does not support the results of Scholz (1968) that the m-value (or the b-value in Gutenberg–Richter's formula) decreases with the increase of applied stress.

Figure 6.27 shows that the magnitude–frequency relation of AE events may be influenced by temperature (T). The specimen used is the cylindrical pine resin which contains dense granular pumice particles, tested at different temperatures. Since the ductility of the medium increases with the increase of temperature, this graph suggests the effect of ductility on the magnitude–frequency relation of AE events. This problem may be not simple and needs further studies.

From the above-mentioned results, it may be concluded that the magnitude–frequency relations of AE events satisfy the Ishimoto-Iida's equation in general, and the relation is strongly dependent on the mechanical structure, such as the degree of

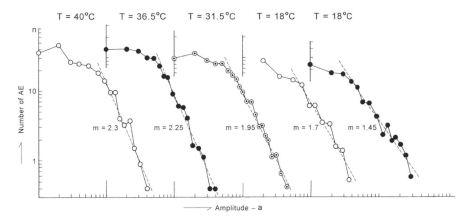

Figure 6.27. The magnitude–frequency relation of AE events in the pine resin specimen which contains densely pumice particles, for different temperatures. The ductility increases with the increase of temperature.

heterogeneity. Further research on the other factors, such as stress level, stress history and ductility, is needed.

REFERENCES

Asada, T., Z. Suzuki and Y. Tomoda. (1951). Note on the energy and frequency of earthquakes. Bull. Earthquake Res. Inst., Tokyo Univ., **29**, 289–293.

Gutenberg, B. and C. F. Richter. (1942). Earthquake magnitude, intensity, energy, and acceleration. Bull. Seismol. Soc. Am., **32**, 163–191.

Ishimoto, M. and K. Iida. (1939). Observations sur les séismes enregistrés par le microseismographe construit derniérement (1). Bull. Earthquake Res. Inst., Tokyo Univ., **17**, 441–478 (in Japanese).

Mogi, K. (1962 a). Study of elastic shocks caused by the fracture of heterogeneous materials and its relations to earthquake phenomena. Bull. Earthquake Res. Inst., Tokyo Univ., **40**, 125–173.

Mogi, K. (1962 b). Magnitude – frequency relation for elastic shocks accompanying fracture of various materials and some related problems in earthquakes. Bull. Earthquake Res. Inst., Tokyo Univ., **40**, 831–853.

Mogi, K. (1980). Amplitude – frequency relationship of acoustic emission events. Program and Abstracts, Seismol. Soc. Japan, 1980 (1), p.176 (in Japanese).

Mogi, K. (1981). Earthquake prediction program in Japan. In "Earthquake Prediction" Maurice Ewing Series **IV** (D. W. Simpson and P. G. Richards, eds.), Am. Geophys. Union, Washington, D. C., 635–666.

Richer, C. F. (1935). An instrumental earthquake magnitude scale, Bull. Seismol. Soc. Am., **25**, 1–32.

Scholz, C. H. (1968). The frequency – magnitude relation of micro fracturing in rock and its relation to earthquakes. Bull. Seismol. Soc. Am., **58**, 399–415.

Tomoda, Y. (1956). Models in statistical seismology, Zisin [ii], **8**, 196–204 (in Japanese).

Vinogradov, S. D. (1959). On the distribution of the number of fractures in dependence on the energy liberated by the destruction of rocks. Izvestiya Akad. Nauk, SSSR, Ser. Geofiz. 1850–1852.

CHAPTER 7

AE activity under cyclic loading

7.1 EFFECT OF TIDAL LOADING

7.1.a *Introduction*

The Izu Peninsula is a resort area located about one hundred kilometers southwest of Tokyo, Japan. In the last twenty years, a number of earthquake swarms have occurred in succession off the east coast of the Izu Peninsula. Since these earthquakes occurred under the sea, this seismic area was affected by cyclic tidal loading. Although there are many reports about the relation between seismic activity and the oceanic tide, the relation is not clear in many cases. However, I found a noticeable example of the tidal effect in the case of the 1980 earthquake swarm off the east coast of the Izu Peninsula.

Fortunately we (Mogi and Mochizuki, 1983) conducted the observations of high-frequency seismic waves (AE events) directly above the focal region of the 1980 earthquake swarm, including a major shock of M 6.7. Although this topic has no direct relation to the effects of the tidal loading, I shall explain our results which are quite rare AE observations, as an introductory topic.

7.1.b *Observations of AE events directly above the focal region of the 1980 earthquake swarm*

In 1980 June–July, a marked earthquake swarm occurred off the east coast of the Izu Peninsula, as an extension of recent seismic activity. The activity gradually increased and the main shock of M 6.7 occurred on 29 June. This earthquake caused some damage including small land slides on the eastern coast of the peninsula. This seismic activity continued for a month, then terminated.

Immediately after the earthquake swarm was initially reported, we decided to try to detect the high frequency seismic waves (AE events). With the cooperation of the Hydrographic Department of the Maritime Safety Agency, Japan, we departed for the sea area where the earthquake swarm occured from Tokyo Bay at noon 28 June on the surveyor ship, Takuyo (Fig. 7.1). The observation using a hydrophone started on the evening of 28 June, and continued for nearly two days until the morning of 30 June. After this earthquake swarm, we started a continuous observation by ocean-bottom hydrophones with a fixed cable system in this region. The results of the AE measurements by the surveyor ship are discussed below.

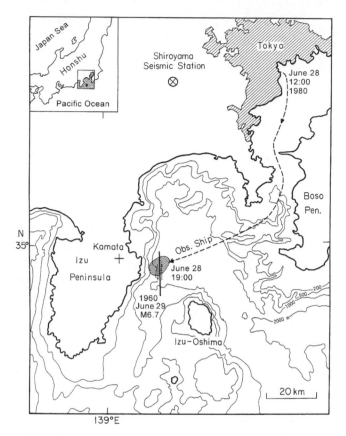

Figure 7.1.   Locality of the strike–slip fault of the Izu-Hanto-toho-oki earthquake on 19 June, 1980 and the submarine topography. Dotted region covers the high frequency seismic wave observation area. Broken curve shows the course of the surveyor ship from Tokyo harbor. ⊗: Shiroyama seismic station of ERI; +: Kamata seismic station of JMA. (Mogi and Mochizuki, 1983).

Figure 7.2 shows the daily changes in the number of earthquakes in this earth-quake swarm that were observed by JMA at Kamata, and the black bars indicate felt earthquakes. The figure also shows the large M 6.7 earthquake and the period of this survey. The activity was extremely strong, and even though our survey only covered 2 days, it took place at the peak of this activity. However, due to bad weather for ~10 hr from the morning of 29 June until the occurrence of the large M 6.7 earthquake at 4:20 pm that day, we had trouble with the coiling of the hydrophone cable around the hull, and observations had to be suspended. We were able to resume observations ~10 min after the main shock, however, and recorded high aftershock activity.

Figure 7.3 shows the tracking chart after observations were resumed. We continued to cruise as slowly as possible within the area I designated earlier. The ship's position was determined about every half hour, with an accuracy of 100 m. Figure 7.4 shows in detail the temporal transition of seismic activity during the period in which we were

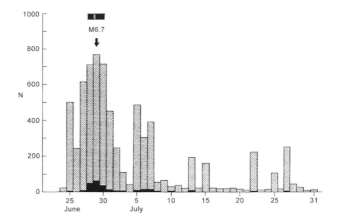

Figure 7.2. Seismic activity around the time of the 1980 Izu-Hanto-toho-oki earthquake (M 6.7) and the duration of the marine observations of high-frequency earthquakes. Vertical axis, number N of daily earthquakes observed at Kamata (JMA); vertical black bars, felt earthquakes; thick horizontal bar, period of hydrophone observations at sea (hatched portion, interrupted). (Mogi and Mochizuki, 1983).

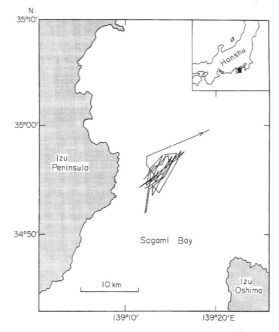

Figure 7.3. Course of the observation vessel immediately after the Izu-Hanto-toho-oki earthquake (M 6.7). (Mogi and Mochizuki, 1983).

carrying out observations on board, and the number of earthquakes that were observed every 10 min at the Shiroyama seismographic station (Earthquake Research Institute), ~70 km to the north. The steadily declining activity continuing from the main shock (M 6.7) represents aftershocks. Peaklike activity lasting ~2 hr overlaps with this, and

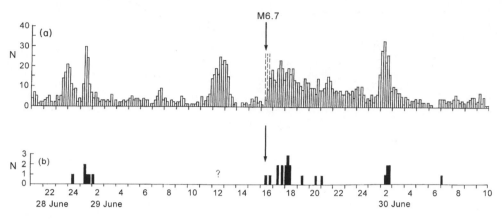

Figure 7.4.   (a) Number of earthquakes observed at 10-min intervals at the Shiroyama seismographic station of the Earthquake Research Institute located ∼70 km north of the main shock (M 6.7, arrow) during the period observations were being carried out at sea. (b) Sea shocks felt aboard the observation vessel. Times are shown in terms of a 24-hr clock. (Mogi and Mochizuki, 1983).

is one of the burst-type earthquake swarms. Figure 7.4b shows the number of shocks felt on board the ship. At the time of the M 6.7 earthquake the ship was thrust up and shaken from side to side, and creaking could be heard all over the ship. The author was up on the bridge watching the eastern coast of the Izu Peninsula. Simultaneously with the earthquake, small-scale landslides occurred and the clouds of dust looked like a tiny eruption. I believe that there are probably no other cases in which an earthquake researcher has personally experienced such a large submarine earthquake directly above its focal region and then felt numerous thrusting-like submarine earthquakes.

Figure 7.5 shows the waveforms of high-frequency earthquakes observed by the hydrophone. A 50–100 Hz filter was used. Earthquakes with various different waveforms were observed, but here I have focused on the sharpness of the onset and the duration of the vibrations and classified them into three types: A, B, and C. Type A represents those with a sudden onset and relatively short duration of vibration, B represents those with a rather slow onset and long duration, and C represents those that had an even slower onset and whose maximum amplitude is not obvious.

The reason for making this classification is that the sharpness of this onset and the short duration directly indicate the distance from the place where observations are carried out. In general, earthquakes of type A are the closest and are observed with little attenuation, so these are thought to have occurred in an extremely shallow location directly beneath the observation ship. Those of type C are distant earthquakes and correspond closely to those observed by distant seismometers on land. Earthquakes of type B lie between the other two types.

Figure 7.6 shows the ship's route from 10 min after the main shock occurred and the three types (A, B, and C) of high-frequency earthquakes observed at this time. The ship's location was determined within an accuracy of 100 m, as mentioned above.

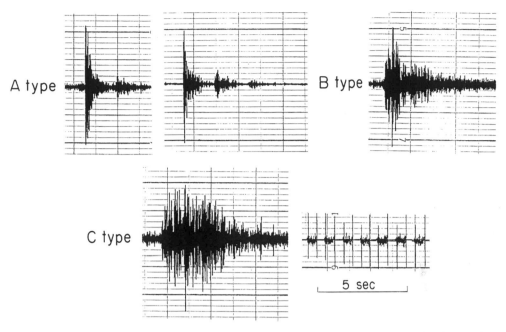

Figure 7.5. Waveforms of the high-frequency earthquakes recorded by the hydrophone, classified into the three types (A, B, and C) according to the sharpness of their onset and the duration of seismic waves. The lower right part of the figure shows the time scale. (Mogi and Mochizuki, 1983).

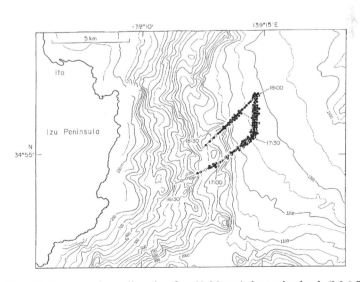

Figure 7.6. The ship's course from directly after (4:30 pm) the main shock (M 6.7) at 4:20 pm 29 June until 6:30 pm, and the three types (●: A, ◐: B, ○: C) of high-frequency earthquakes observed at that location. (Mogi and Mochizuki, 1983).

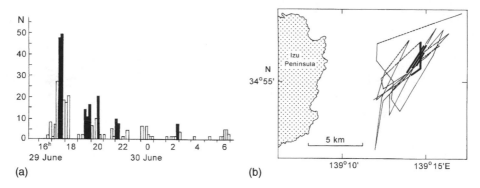

Figure 7.7.   (a) Temporal changes in the observed number of A-type high-frequency earth-quakes (time shown in terms of a 24-hr clock). Peaks of activity are shown in black. (b) The area where the peaks of activity in part (a) were observed is indicated by thick lines, showing the course of the observation vessel during the same period. (Mogi and Mochizuki, 1983).

Since A-type earthquakes are assumed to have occurred directly beneath the observation vessel, their locations along the ship's course can be considered as the locations of the hypocenters. As Fig. 7.6 shows, when the main shock occurred, the ship was traveling ~4 km from the eastern coast of the Izu Peninsula, and mostly C-type earthquakes were observed. As the vessel moved northeastward, B-type and then A-type earthquakes were observed, and ~8 km from the eastern coast of the peninsula an extremely large number of A-type high-frequency vibrations were observed. Moving to the southwest again, the number of A-type earthquakes decreased, and C-type earthquakes were observed.

Figure 7.7a shows the temporal changes in the number of A-type high-frequency earthquakes after the main shock, and it is noteworthy that whereas the normal seismic activity observed at a distance declined steadily after the main shock, as can be seen in Fig. 7.4a, these reached a marked peak and then gradually declined. Particularly observed marked peaks are blackened. In Fig. 7.7b the places where numerous A-type earthquakes were observed while the observation vessel went back and forth, are shown with thick lines and reveal how A-type high-frequency vibrations always increased sharply when passing over certain locations. This concentration of A-type high-frequency vibrations in a certain location indicates that the seismic fault passes through this location and that the fault is shallow, reaching up as far as the ocean floor. It was inferred from an analysis of seismic observation data on land that the seismic fault ran in a north-south direction and was a 20 km-long left lateral strike-slip type. From the results shown in Fig. 7.7, it is estimated that it was located ~8 km from the eastern coast of the Izu Peninsula. The location of this seismic fault is shown in Fig. 7.8 by the stippled belt. This figure also shows outlines of the aftershock region of the M 6.7 earthquake that were reported by various research organizations and researchers. Such striking discrepancies among various research groups arise from the locations of observation points and the assumed seismic velocity structures.

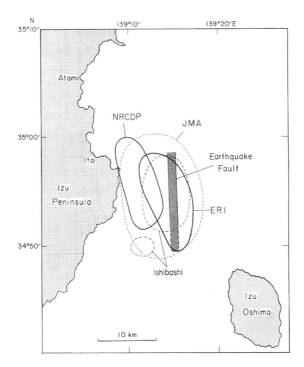

Figure 7.8. The stippled belt shows the location of the seismic fault inferred by collating seismic observation results obtained on land and the results of on-board observations of high-frequency earthquakes (Mogi and Mochizuki, 1983). The ovals show the outlines of aftershock regions found by various organizations, based on their respective seismic observation networks. Aftershock regions are based on data from ERI, Earthquake Research Institute (Karakama et al., 1980); JMA, Japan Meteorological Agency (1980); NRCDP, National Research Center for Disaster Prevention (Ohtake et al., 1980, and Ishibashi (1980)). (Mogi and Mochizuki, 1983).

At that time, the distribution of aftershocks obtained from a wide seismic network was the sole basis for estimating the location of submarine seismic faults. However, since seismic stations were located only on land, and the information on the velocity distribution in vertical and horizontal directions within the earth's crust was insufficient, the accuracy in the determination of the hypocentral distribution of aftershocks was low. In contrast, by focusing on the waveforms of high-frequency vibrations and on the spatial distribution of the degree of activity of A-type earthquakes, which show a sharp onset and short duration, it was possible to determine the location objectively, rapidly, and with high accuracy, as discussed above. In this case, only errors in determining the location of the observation vessel affect the determination of the fault's location, and as stated above, the determination of the ship's location was extremely accurate, to within 100 m. However, it is difficult to routinely implement such a method normally.

This result can be verified from independent data. Figure 7.9 shows two different waveforms. One is a simple single-type waveform, while three or four phases at equal

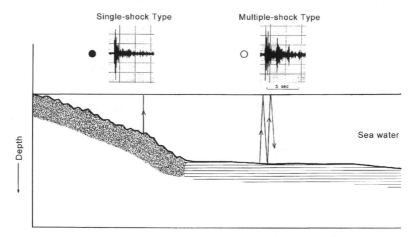

Figure 7.9.   Waveforms of the high-frequency earthquakes observed by hydrophone, and the state of the sea bottom inferred from these waveforms. (Mogi and Mochizuki, 1983).

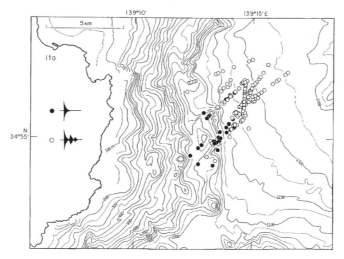

Figure 7.10.   Spatial distribution of simple-shock-type vibration (●) and multiple-reflection-type vibrations (○). (Mogi and Mochizuki, 1983).

intervals can be recognized in the other, and their amplitudes decrease systematically. Such differences in the waveforms are thought to be due to the existence or nonexistence of a reflection phase caused by differences in the surface of the ocean bottom. Simple-type waveforms occur when the surface of the ocean bottom is extremely uneven and there are almost no reflections from the surface, while the multiple-reflection type occurs when the ocean bottom is flat and the degree of reflection is high. A spatial distribution of the simple type (●) and the multiple-reflection type (○) is shown in Fig. 7.10. The location of the open and solid circles are separated, with the solid circles distributed in extremely uneven places on the western slope of this sea area, and the open circles distributed in the flat part of the offing to the east. The distribution of these open and solid circles coincide on the whole with the submarine topographical map and indicates that the error in estimating the location of the focal regions of the A-type earthquakes is not large. At the same time, most A-type earthquakes are of the multiple-reflection type, showing that the seismic fault reaches to the surface of the flat part of the seafloor, and it seems to be somewhat east of the location of the active fault that has been estimated so far from submarine topography.

The results of observing high-frequency earthquakes also provide important information about the depth of the fault place. According to preliminary reports on aftershock observations by the earthquake observation network of several institutions, the aftershocks were quite deep (10–20 km). As stated above, however, a large number of A-type high-frequency earthquakes (AE events) thought to have originated in an extremely shallow part of the seafloor were observed, showing that the fault is shallow, almost reaching the seafloor. This cannot be explained by assuming a fault plane 10–20 km deep. Matsu'ura (1983) carefully redetermined the three-dimensional locations of the aftershocks and obtained a shallow value (0–10 km). In the cases of the 1974 Izu-Hanto-oki earthquake (M 6.9), and the 1978 Izu-Oshima-kinkai earthquake (M 7.0) which occurred in the Izu Peninsula region, movements of the fault were observed on the land surface, and it is very natural that the seismic fault of the 1980 Izu-Hanto-toho-oki earthquake, which was associated with this event, was also shallow, reaching to the surface of the sea floor.

Collating these results, the sequence shown in Fig. 7.11 is assumed to be the process by which the Izu-Hanto-toho-oki earthquake occurred. First, earthquake swarms began to occur about 25 June at a site several kilometers off shore from the eastern coast of the Izu Peninsula (Fig. 7.11 a). Since hydrophone observations from the night of June 28th until the morning of June 29th failed to observe any A-type high-frequency vibrations, it is assumed that these earthquake swarms were relatively deep. (Due to the high noise level of these hydrophone observations, it was not possible on this occasion to ascertain whether or not shallow high-frequency microvibrations occurred prior to the large earthquake.) As the days passed, this earthquake swarm activity increased, and a large-scale fracture grew rapidly from this active part and reached as far as the surface of the seafloor, over a total length of 20 km. This was the large M 6.7 earthquake (Fig. 7.11b). Aftershocks then continued to occur on and around the fault plane, but their activity systematically decreased as time passed. However, burst-type earthquake swarms continued to occur intermittently up until the end of July, while gradually becoming less active. It is estimated from the focal mechanism that the crustal stress that caused this fracture was compression in a northwest-southeast direction (or tension in a northeast-southwest direction).

7.1.c *Seismic activity and ocean tide*

Figure 7.12 shows the temporal variation of the number of earthquakes (M ≥ 2.0) in the 1980 Izu-hanto-oki earthquake swarm. This was obtained by the observation at

272   *Experimental rock mechanics*

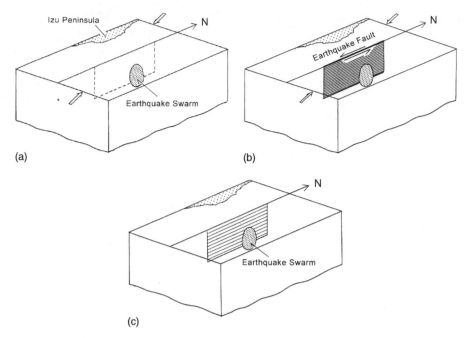

Figure 7.11.   Schematic illustration showing the process by which the 1980 Izu-Hanto-toho-oki earthquake occurred. (a) Precursory earthquake swarms. (b) Main shock (M 6.7). (c) Gradually abating earthquake swarms. (Mogi and Mochizuki, 1983).

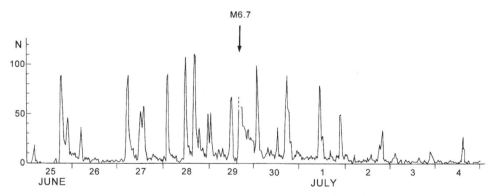

Figure 7.12.   The temporal variation of number of earthquakes in the 1980 Izu-Hanto-toho-oki earthquake swarm observed at the seismographic station of the Earthquake Research Institute, University of Tokyo, ~70 km to the north. Arrow: the 1980 Izu-Hanto-toho-oki earthquake of M 6.7.

"Shiroyama" seismographic station of the Earthquake Research Institute, University of Tokyo, ~70 km to the north (see Fig. 7.1). The arrow indicates the 1980 Izu-hanto-oki earthquake of M 6.7, which is discussed in the preceding section.

Figure 7.13 shows the M-T graph of the 1980 earthquake swarm. The period of this swarm activity was divided into three periods, A, B and C. In the period B from June

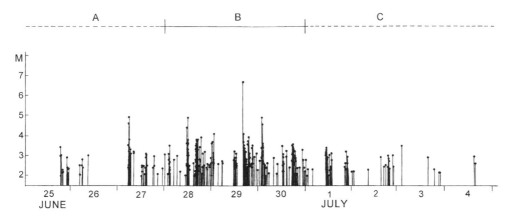

Figure 7.13. M–T graph of the 1980 earthquake swarm. A close relationship between the seismic activity and oceanic tide was found in the period B in which the seismic activity was high.

28 to 30, the seismic activity was highest and the M 6.7 earthquake occurred. In this period, the present author found a noticeable relation between seismic activity and the oceanic tide (Mogi, 1983).

In Figs. 7.14–7.16, the relation between the oceanic tide and the seismic activity in the period B (June 28, 29 and 30) are shown, respectively. In these figures, the top is the temporal variation of the sea level (H), the middle is the number of earthquakes (M ≥ 2.0) in this swarm region and the bottom is the M–T (magnitude–time) graph for this seismic activity. In these figures, it was noticed that the temporal variations of seismic activity in these three days had some similarities.

To examine this problem further, the oceanic tidal curves and the frequency curves of earthquakes in these three days are shown together in Fig. 7.17. In this figure, it seems that the shift of the peaks of the seismic activity to the right is nearly proportional to that of the tidal curves. Figure 7.18 shows the temporal variations of the sea level and the number of earthquakes on June 28, 29 and 30, in which the seismic frequency curves are shifted on the basis of the daily phase shift of the oceanic tidal curve. Figure 7.19 shows the relation between the oceanic tidal curves and the seismic activity illustrated by the M–T graph. The procedure for this graph is the same as Fig. 7.18.

Figures 7.18 and 19 strongly support that there was a marked relation between the seismic activity and the oceanic tidal loading. Furthermore, if we accept this result, it may be concluded that the occurrence of the 1980 earthquake of M 6.7 was effected or triggered by the oceanic tidal loading. In Fig. 7.19, we can see a similar regular pattern in the M–T graph for June 28, 29 and 30, and that the major earthquake of M 6.7 occurred as a member of this regular pattern. This is a quite interesting result.

In the above discussion, the activity of only period B in Fig. 7.13 was treated. We examined this subject in period A and period C. Figure 7.20 (a) and (b) show the temporal variation of the number of earthquakes in period A (June 25–27) and period C (July 1–4), respectively. In these figures, we cannot find any correlations.

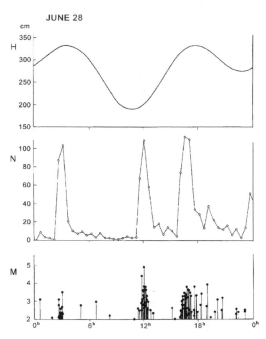

Figure 7.14.   Relation between the oceanic tide and the seismic activity in June 28, 1980. H: Sea level, N: number of earthquakes, M: earthquake magnitude. (Mogi, 1983).

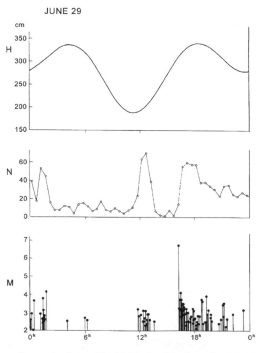

Figure 7.15.   The same figure in June 29, 1980. In this day, the M 6.7 earthquake occurred.

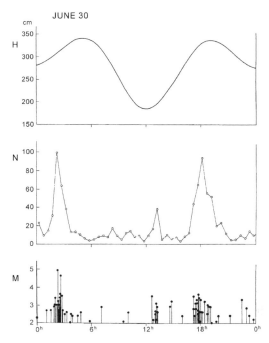

Figure 7.16. The same figure in June 30, 1980.

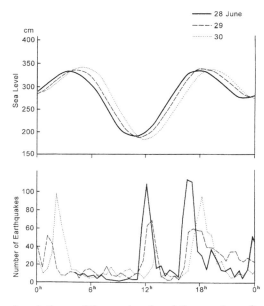

Figure 7.17. Temporal variations of the sea level and the number of earthquakes in June 28, 29 and 30.

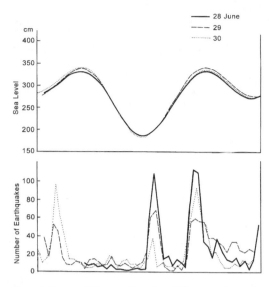

Figure 7.18.   Temporal variations of the sea level and the number of earthquakes in June 28, 29 and 30, in which curves of number of earthquakes are shifted on the basis of the daily shift of the tide curve.

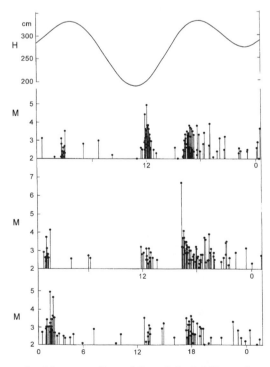

Figure 7.19.   The oceanic tide curve (June 28) and the M–T graphs of June 28, 29 and 30. The procedure is the same as Fig. 7.18. (Mogi, 1983).

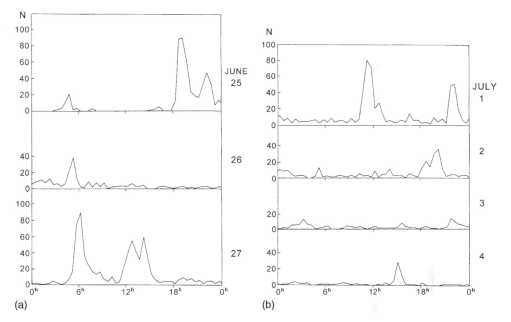

Figure 7.20. (a) The temporal variation of number of earthquakes in the period A (June 25–27). (b) The temporal variation of number of earthquakes in the period C (July 1–4).

The most noticeable difference between period B and the other periods (A and C) is the difference of the degree of seismic activity.

Figure 7.21 shows some examples of a compression test of a granite specimen in the laboratory. σ is the applied stress, ε is the strain and n is the number of AE events. This tested granite sample was partially ruptured and under an unstable condition. In such a case, the marked increase of AE activity can occur by the increase of a small fraction of applied stress. The stress difference caused by the oceanic tide is very small, and so generally the seismic activity is not influenced significantly by the oceanic tide. However, in the special case in which the mechanical state in a region is particularly complex and unstable, a small fraction of the stress, such as the oceanic tide, may be able to cause a significant effect on the seismic activity. As seen in Fig. 7.11, it is probable that the focal region of the 1980 earthquake was mechanically complex and so it may have been unstable during the few days before and after the M 6.7 earthquake. This special condition in this region may explain the marked effect of the tidal loading on the seismic activity.

7.2   AE UNDER CYCLIC COMPRESSION

7.2.a   *Experimental procedure*

As mentioned above, the AE activity of heterogeneous brittle rocks increases monotonically with increasing stress and generally depends on stress level.

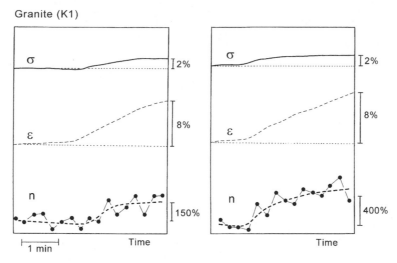

Figure 7.21.   Examples of the case in which the number of AE events markedly increases by a fraction of the applied stress, in laboratory experiments.

However, Kaiser (1953) found that AE activity shows a marked stress history effect. That is, the AE activity increases with the increasing stress, but in the case of reloading, it is appreciably low and begins to increase markedly at the previously applied maximum stress level. Based on this effect, the previously applied maximum stress can be estimated from the curve of the AE activity under monotonically increasing stress. This effect has been observed in industrial materials, and also, the Kaiser effect has been observed for rocks (e.g. Kaiser, 1953; Goodman, 1963).

In this section, the AE activity under cyclic uniaxial compression observed in laboratory experiments and some noticeable features of AE phenomena obtained from the experiments are mentioned (Yoshikawa and Mogi, 1981; 1990).

Figure 7.22 shows the main part of the experimental method, and the specimen design, which is similar to Mogi (1966). The rock specimen is a right circular cylinder ϕ 20 mm $\times$ 60 mm, or ϕ 30 mm $\times$ 90 mm connected to steel end pieces by a steel epoxy fillet. The gradual decrease of thickness of the epoxy fillet eliminates most of the stress concentration, as is explained below (also see Chapter 1).

In Figs. 7.23a and 23b, a finite element method is applied to a three dimensional axisymmetrical model (e.g. Zienkiewicz, 1971) to find the stress distribution within the cylinder. First for the case when stress is applied directly to the rock specimen through the steel end pieces, and secondly when the epoxy fillets are connected to the specimen. When a rock sample is directly compressed by steel end pieces, a very large concentration of stress arises at the corner of the specimen, as seen in Fig. 7.23a. This is due to the generation of a shear stress along the contact plane because of the difference in elasticity between rock and steel, as well as the sudden change of cross sectional area at the contact of the two. As shown in Fig. 7.23b, when the epoxy fillet is attached, the stress level gradually decreases from the center of the specimen to the

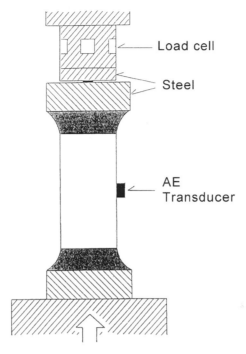

Figure 7.22.   Setup of AE measurements in the rock sample under cyclic uniaxial compression. (Yoshikawa and Mogi, 1981, 1990).

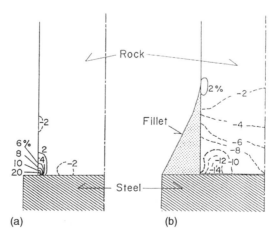

Figure 7.23.   Calculated axial stress distribution, shown by deviation from the mean applied stress in the middle of the specimen. Positive values indicate excess and negative values indicate deficit of stress values normalized by the mean applied stress, respectively. (a) Direct application of load to rock. (b) Application of load from steel to rock with fillet. Young's modulus (in GPa) and Poisson's ratio for rock, steel or fillet are 30 and 0.15 (rock), 200 and 0.30 (steel) and 3 and 0.4 (fillet), respectively. (Yoshikawa and Mogi, 1990).

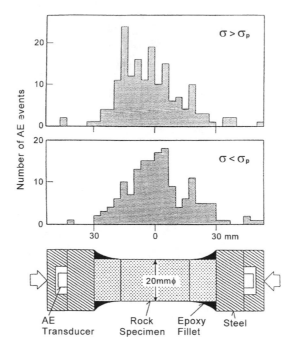

Figure 7.24.   Uniaxial spatial distribution of AE sources, determined by the time difference between the first motions of each AE signals received at the transducers coupled to the specimen ends. Top: at stresses higher than the previous maximum stress. Center: at stresses lower than the previous maximum stress. Bottom: design of test specimen. (Yoshikawa and Mogi, 1990).

end. In this case, the concentration of stress at the end of the rock specimen becomes negligible. By adopting this design, we can apply a uniform compressive stress to the central main part of the specimen.

The lack of stress concentrations in Fig. 7.23b was verified by a measurement of the AE source distribution. Figure 7.24 shows the uniaxial spatial distribution of the AE sources measured when stress is applied to a specimen with epoxy fillets: the source distribution is shown at two different compression levels. The first is a compressive stress higher than the previously applied maximum stress (σ_p); the second is lower than σ_p. In both cases, the AE events do not concentrate at the ends of the specimen, most AE events occur in the central main portion of the specimen. These AE source locations indicate the nearly uniform stress distribution in the central main portion of the specimen and the elimination of stress concentrations at the ends due to the epoxy fillet.

7.2.b   *AE events under cyclic compression*

Figure 7.25 shows the AE activity for the initial loading and reloading, and stress as functions of time in Shinkomatsu andesite. Shinkomatsu andesite has a porosity of

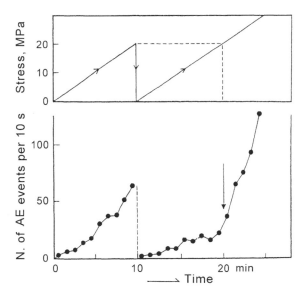

Figure 7.25.   AE activity for the initial loading and reloading, and stress as functions of time in Shinkomatsu andesite. Top figure: AE activity (the solid circles are the number (N) of AE events per 10 s) (Yoshikawa and Mogi, 1981).

5.4% and a heterogeneous structure, so the AE activity is relatively high even at low stress levels. The uniaxial compressive strength is about 100 MPa. In this figure, the solid circles in the AE activity curve show the number of AE events per 10 s.

In this figure, a virgin rock sample is initially loaded to a stress value (σ_p) and then unloaded. After this process, the sample is reloaded monotonically. The AE activity for this reloading begins to increases markedly at the previous maximum stress level. Therefore, we can estimate the previous maximum stress from the change in the slope of the AE activity curve under monotonically increasing stress. This effect on the AE activity by the stress history is "the Kaiser effect" and the stress estimated on the basis of this effect is designated as σ_e in this section.

Figure 7.26 shows the AE activity curves for successive loadings as functions of the applied uniaxial compressive stress. The maximum stress level in each loading increases successively. Different symbols indicate the AE activity curves for each loading. In these experimental results, it was found that the stress value (σ_e) estimated on the basis of the Kaiser effect does not always agree with the previously applied stress (σ_p).

Figure 7.27 shows the difference between the previously applied stress (σ_p) and the stress (σ_e) estimated by the Kaiser effect. In this figure, the vertical axis and horizontal axis are normalized by the uniaxial compressive strength (σ_f). Different symbols show different conditions, such as dry or wet, during the experiments (Yoshikawa and Mogi, 1981). According to these experimental results, the stress estimated by the Kaiser effect is lower than the previously applied stress and the difference ($\sigma_p - \sigma_e)/\sigma_f$

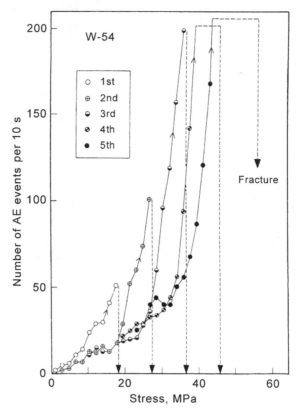

Figure 7.26.　AE count rate for the cyclic loadings as functions of stress. The maximum stress level in each loading increases successively (Yoshikawa and Mogi, 1981).

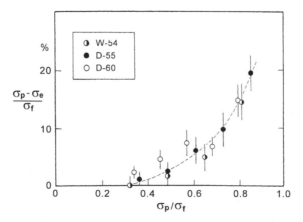

Figure 7.27.　Difference between the previously applied stress (σ_p) and the stress (σ_e) estimated by the Kaiser effect, as a function of the applied stress. σ_f: uniaxial compressive strength (Yoshikawa and Mogi, 1981).

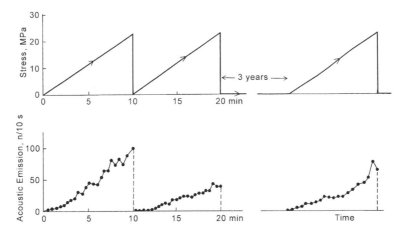

Figure 7.28. Example of the AE activity curve measured from initial loading through reloading. Top: temporal variation of stress; bottom: AE activity for the first, second and the last loadings. Solid circle: number of AE events per 10 s (Yoshikawa and Mogi, 1990).

increases with the increase of σ_p/σ_f. The difference is very small at low stress levels ($\sigma_p/\sigma_f < 0.4$).

The time dependence of the Kaiser effect is an important problem. In order to measure the time dependence, following to the initial loading, the rock specimens were held at room temperature under either dry or water-saturated conditions. Each specimen was then reloaded and the AE activity was measured. At the same time, the "initial stress" was estimated, as mentioned above. The duration of the hold (T) from the initial loading to reloading was varied from a few minutes to 894 days. An example of the AE activity curves measured from the initial loading through reloading is shown in Fig. 7.28. When a sample was reloaded just after the initial loading, the AE activity was much lower than that of the initial loading. In this case, the initial stress can be estimated with high accuracy. However, when the duration (T) was extended to 3 years, a considerable recovery of the AE activity occurred, as shown in Fig. 7.28. In such cases, the estimated stress (σ_e) is significantly lower than the initial stress value.

Figure 7.29 shows the difference between the initial stress (σ_p) and the estimated stress (σ_e) as functions of elapsed time (T). Different symbols indicate the different levels of the initially applied stress (σ_p). The above results prove that the Kaiser effect is dependent on time. The higher the initial stress level, the larger the time dependence.

Figure 7.30 shows the AE activity curve observed under repeated loading. After an initial stress of 25.5 MPa is applied, an additional maximum stress of up to 42.4 MPa is repeatedly applied at a constant rate of 53 kPa/s. At the beginning of the repeated loading (or the second loading) the AE activity increases rapidly at the maximum value of the initial stress (the Kaiser effect). However, from the next loading, the AE activity curve repeats with nearly the same pattern of change, irrespective of the initial stress.

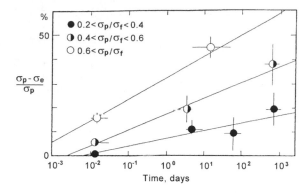

Figure 7.29.   Difference in the initial stress value (σ_p) and the estimated stress value (σ_e) as a function of elapsed time from initial loading to reloading for different stress levels (σ_p/σ_f). Each plot represents the average values (Yoshikawa and Mogi, 1990).

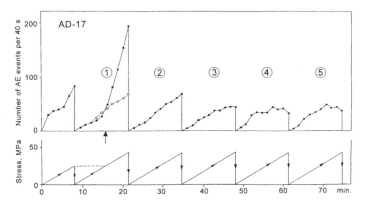

Figure 7.30.   AE activity for cyclic loading as a function of time. Solid circles: number of AE events per 40 s; open circles: number of AE events of the subsequent cycle. AE activities during unloading processes are omitted (Yoshikawa and Mogi, 1990).

This result suggests that two types of AE events occur under monotonically increasing stress. Type I AE events begin to occur when the applied stress passes the previously applied maximum stress. Type II AE events occur even at a stress level lower than the previous one, and show a similar activity curve for each loading cycle. The source locations of both types of AE events are shown in Fig. 7.24. They occur in the central portion of the specimens, with no significant difference in spatial distribution.

In Fig. 7.31, the AE activity curves under repeated loading to the maximum stress level of 55 MPa, at a rate of 46 kPa/s are shown. The top figure shows curves from the 3rd to the 11th loading and the bottom figure shows curves from the 52th to the 60th loading.

It is reasonable to conclude that Type I AE activity in this rock is associated with the formation of stress-induced microfractures since the activity increases markedly

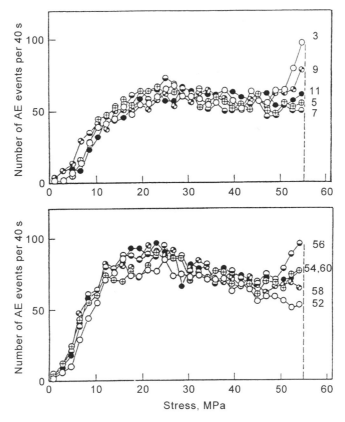

Figure 7.31.   AE activities for cyclic loadings as functions of stress. The upper figure shows curves from the 3rd to the 11th loading. The lower figure shows curves from the 52nd to the 60th loading (Yoshikawa and Mogi, 1990).

with an increment of stress beyond the previous maximum stress. On the other hand, various features of Type II AE may be attributed to the frictional sliding along existing cracks.

The above-mentioned experimental results are summarized as follows:

1. There are generally two types of AE events which occur in rocks when uniaxial compression is applied: Type I AE appears when the maximum previous stress is exceeded. Type II AE appears at stress levels below the maximum previous stress.
2. Type I AE occurs due to the formation of new microfractures. The Kaiser effect can be observed during the onset of Type I AE activity. Type II AE activity is obtained by repeated loading.
3. The time dependence of the Kaiser effect was obtained. The stress value estimated by using the Kaiser effect decreased as time elapses; in particular, the higher the initial stress level, the lower the estimated stress value.

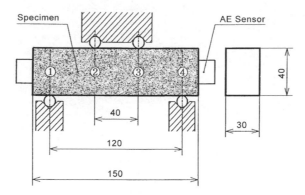

Figure 7.32.   Schematic view of the four-point bending test arrangement are indicated in mm (Mogi, Mori and Saruhashi, 1983–1994 see text).

7.3   AE UNDER CYCLIC BENDING

7.3.a   *Introduction*

As mentioned in 7.1.c, there are sometimes noticeable relations between seismic activity and the oceanic tidal loading. Particularly the results shown in Fig. 7.18 and 7.19 are significant. To clarify the mechanism of the effect of repeated loadings, we measured the temporal variation of AE activity of rocks under cyclic bending. In these experiments, particularly a great number of cyclic bending were applied. In this section gradual variations of AE activity and source locations are shown as functions of the number of loading cycles.

The schematic view of the cyclic bending experiment is shown in Fig. 7.32. The tested rock was a granodiorite. The parallelepiped specimen, with dimensions shown by numerals in this figure, was loaded in a four-point bending frame which was installed on a servo-controlled hydraulic testing machine. The loading of sinusoidal waveforms at a frequency of 0.4 Hz was applied. The amplitudes of acoustic waves and source locations of AE events for each loading were measured by the methods described in the previous sections.

We carried out cyclic bending experiments of two types, **A** and **B**, in which the loading processes were different (Mogi and Mori, 1993; Saruhashi, Zeng, Mori and Mogi, 1993; Mori, Saruhashi and Mogi, 1994; Saruhashi, 1994). These experimental results are explained below.

7.3.b   *Experiment A*

Figure 7.33 schematically shows the process of cyclic loading. The vertical axis shows the applied load P (kN) and the horizontal axis shows the number of cyclic loads with a frequency of 0.4 Hz. The amplitudes of cyclic loads were increased at successive intervals until macroscopic fracture of the rock specimen occurred. The

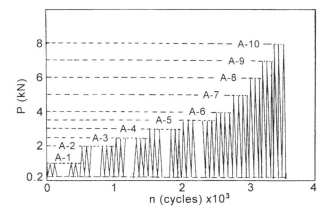

Figure 7.33.   Schematic diagram of cyclic loadings in Experiment **A**. The vertical axis shows the applied load (kN) and the horizontal axis shows the number of cyclic loadings. The maximum loads are increased successively with some intervals and the minimum stress is kept at 0.2 kN. The temporal variation of a cyclic load is of a sinusoidal type.

minimum stress level is kept at 0.2 kN. The temporal variation of a cyclic load is a sinusoidal curve.

In Fig. 7.34 (a), the one-dimensional source locations of AE events along the length of the rock specimen ($=$ X axis) are plotted as a function of the number of cyclic loads. The maximum load and the minimum load are 2 kN and 0.2 kN, respectively. Triangles along the vertical axis indicate the four load points for application of the bending stress. In the following, the AE events which occurred in the central part (X $=$ 55–95 mm, which is 40 mm interval shown as ②–③ in Fig. 7.32) are discussed. In this central part, a uniform bending moment was applied, so uniform tensile stress was generated in the lower surface layer. This figure shows that AE events continued to occur at two places, but the AE activity was relatively low in this case.

Figure 7.34 (b) shows source locations of AE events as functions of the number of loading cycles of which the maximum load is 2.5 kN. The AE activity increased clearly in the central part and the number of AE events continued to occur mainly at several places. Figure 7.34 (c) shows the case for a maximum load of 5 kN. The degree of AE activity and the number of AE source regions increased markedly. This figure shows that AE events continued to occur at many of the same places under cyclic bending.

Figure 7.35 shows the time in a load cycle at which AE events occurred, as a function of the number of loading cycles. The vertical axis shows the time in a load cycle and the horizontal axis shows the number of cycles. On the right side, the applied sinusoidal loading curve is shown along the vertical axis.

Figure 7.35 (a) shows the case in which the maximum load and the minimum load are 2 kN and 0.2 kN, respectively. In this case, AE events began to occur at a relatively high load level and this load level gradually decreased with cyclic loadings. Then AE events continued to occur at a constant load level. In the case of the maximum load 2 kN, AE events occurred only in the loading process, but not in the unloading process.

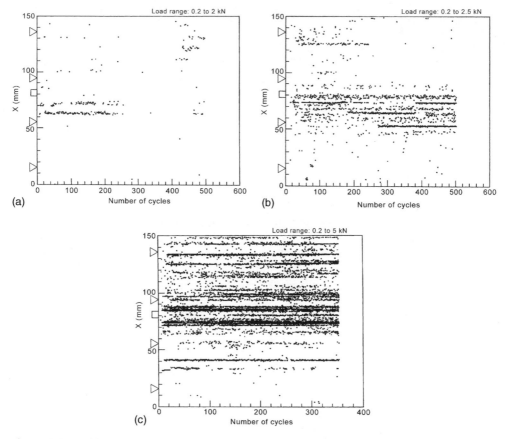

Figure 7.34.   (a) One-dimensional source locations of AE events along the length of rock specimen (= X axis) are plotted as function of the number of loading cycles. The minimum load and the maximum load are 0.2 and 2 kN, respectively. Triangles in the vertical axis (X) indicate the four points for application of the bending load. (b) The same as Fig. 7.34 (a). The maximum load is 2.5 kN. (c) The same as Fig. 7.34 (a). The maximum load is 5 kN.

Figures 7.35 (b) (c) show similar graphs in which the maximum load is 3.5 kN, and 4 kN, respectively. In these cases, the AE activity increased with increasing load level and AE events continued to occur in both the loading and unloading processes. It is remarked that the patterns of the occurrence of AE events are similar in these cases, and that there is no activity around the peak of the loading. However, in the case of the maximum load 6 kN, AE events began to occur around the peak of loading, as shown in Fig. 7.35 (d). After the higher loading, macroscopic fracture occurred.

7.3.c   *Experiment B*

Following Experiment **A**, we carried out experiment **B** which was composed of Experiment B_1 and B_2, to clarify in more detail the effects of cyclic loadings. The experimental procedure is shown in Fig. 7.36. The rock tested was the same

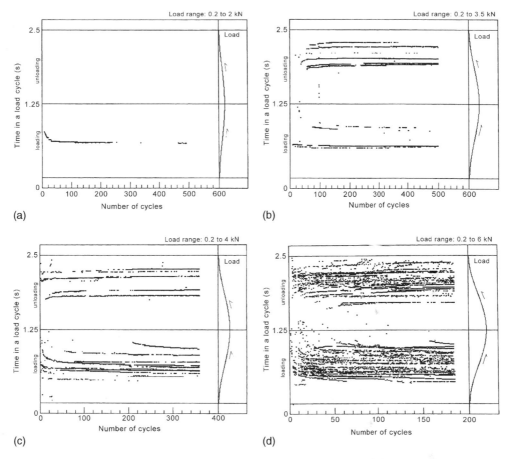

Figure 7.35.   (a) Vertical axis shows the time in a load cycle at which AE events occurred. Horizontal axis shows the number of loading cycles. The sinusoidal temporal variation of loading curve is shown in the right graph. The maximum load is 2 kN. (b) The same as Fig. 7.35(a). The maximum load is 3.5 kN. (c) The same as Fig. 7.35(a). The maximum load is 4 kN. (d) The same as Fig. 7.35(a). The maximum load is 6 kN.

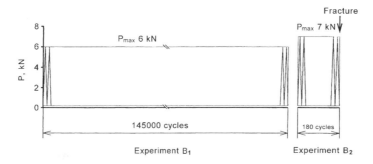

Figure 7.36.   Experimental procedure in Experiment **B**. B_1: the maximum load (P_{max}) is 6 kN and the total number of loading cycles (n) is 145,000; B_2 : P_{max} is 7 kN and n is 180.

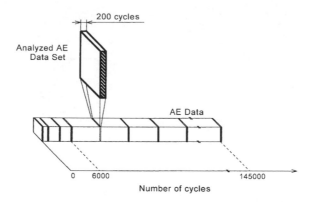

Figure 7.37.   Sampling method of AE data set in AE analysis in Experiment B_1. AE data for each 200 loading cycles were picked up from the whole AE data.

Table 7.1.   List of the analyzed data sets in Experiment B_1

No.	Analyzed cycle	No.	Analyzed cycle
6-1	0 to 200	6-18	69200 to 69400
6-2	1000 to 1200	6-19	76250 to 76450
6-3	2000 to 2200	6-20	79800 to 80000
6-4	4150 to 4350	6-21	87100 to 87300
6-5	6300 to 6500	6-22	90900 to 91100
6-6	9350 to 9550	6-23	94600 to 94800
6-7	15600 to 15800	6-24	102150 to 102350
6-8	21600 to 21800	6-25	106850 to 107050
6-9	24700 to 24900	6-26	110750 to 110950
6-10	31150 to 31350	6-27	114750 to 117950
6-11	34500 to 34700	6-28	122100 to 122300
6-12	41150 to 41350	6-29	125950 to 126150
6-13	44550 to 44750	6-30	129500 to 129700
6-14	50200 to 50400	6-31	137150 to 137350
6-15	55250 to 55450	6-32	141050 to 141250
6-16	62150 to 62350	6-33	145050 to 145250
6-17	65800 to 66000		

granodiorite, but the specimen used in Experiment **B** was not the same as the specimen used in Experiment **A**. In Experiment B_1, the maximum load (P_{max}) was kept at 6 kN and the total number of loading cycles (n) reached 145,000. Following Experiment B_1, the maximum load (P_{max}) was increased to 7 kN and then the main rupture occurred (Experiment B_2), as shown in Fig. 7.36. Just after the 180th cyclic loading, the main fracture occurred. Figure 7.37 shows the sampling method of the AE data set used in the analysis in Experiment B_1. AE data for each of the 200 load cycles were picked from the whole AE data, as shown in this figure. Table 7.1 and 7.2 show the list of the analyzed data sets in Experiments B_1 and B_2, respectively.

Table 7.2. List of the analyzed
data sets in Experiment B_2.

No.	Analyzed Cycle
7-1	0 to 10
7-2	10 to 20
7-3	20 to 30
7-4	30 to 40
7-5	40 to 50
7-6	50 to 60
7-7	60 to 70
7-8	70 to 80
7-9	80 to 90
7-10	90 to 100
7-11	100 to 110
7-12	110 to 120
7-13	120 to 130
7-14	130 to 140
7-15	140 to 150
7-16	150 to 160

Figures 7.38 (a) (b) show AE events plotted graphs where the vertical axis indicates the time in a load cycle and the horizontal axis indicates the number of loading cycles. The sinusoidal temporal variation of the applied load is shown on the right side. The patterns in the initial stage (6-1) and in the latest stage (6-33) are nearly similar.

Figures 7.39 (a) (b) show the AE event count rate as functions of time during a loading cycle for each analyzed data set (No. 6-1–No. 6-22) in Table 7.1. The top figure shows the temporal variation of the applied cyclic load. AE events noticeably occur during both the loading and the unloading, and the pattern of the temporal variation of the AE count rate in each cycle does not change from the initial stage to the later stage.

Next, we examined the temporal change of the energy (E) released as acoustic emission waves under cyclic loading and found a remarkable change in the energy release pattern in the later stage. In Fig. 7.40 (a) (b), $\Sigma\sqrt{E}$ per one loading cycle is shown as a function of time in a load cycle. The patterns are nearly similar in the early stages from No. 6-1 to No. 6-12 but they began to change markedly from No. 6-13. In the loading process, the very large energy release by AE events occurred during a very limited time before the peak load. Also in the unloading process, a fairly large energy release by AE events continued to occur in the limited time, except for No. 6-13–No. 6-17. This result shows that, although the pattern of AE event count rate does not change throughout Experiment B_1, the pattern of energy release by AE events changes markedly from the intermediate stage to the later stage. This result suggests a gradual change of mechanical properties of the rock specimen caused by a great number of loading cycles.

Figure 7.42 shows AE event counts per one loading cycle as functions of the number of cycles. Solid and open circles show values for No. 6-1–No. 6-33 in the loading

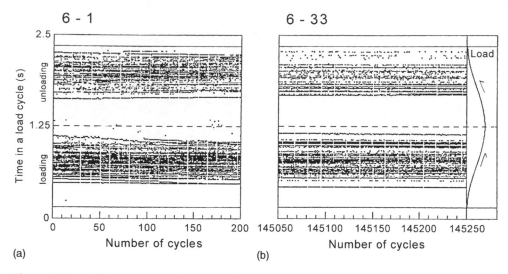

Figure 7.38.   AE events are plotted in the graph in which the vertical axis shows the time in a load circle and the horizontal axis shows the number of loading cycles. The sinusoidal temporal variation of applied load is shown in the right side. (a) Initial stage (No. 6-1 in Table 7.1), (b): final stage (No. 6-33 in Table 7.1) in Experiment B_1.

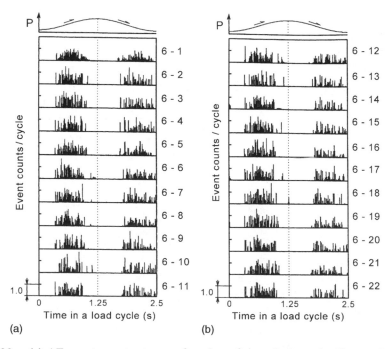

Figure 7.39.   (a) AE event count rate as a function of time during a loading cycle. The top figure shows the temporal variation of load. Numerals on the right side indicate the data set numbers in Table 7.1. (b) Continued.

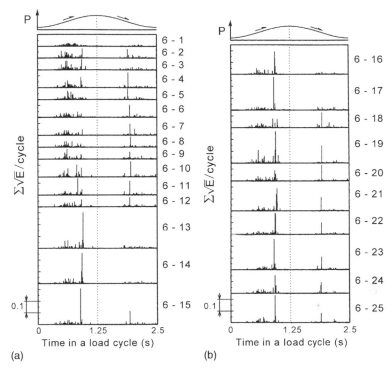

Figure 7.40. (a) AE activity shown by $\Sigma\sqrt{E}$/cycle as a function of time during a load cycle. Numerals on the right side indicate the data set numbers in Table 7.1. E: energy. (b) Continued.

and unloading processes, respectively. The definition of the loading process and the unloading process is explained in Fig. 7.41. According to Fig. 7.42, the AE count rate in the loading process is higher than that in the unloading process, and the activity in the unloading process decreases gradually with increasing number of loading cycles.

Figure 7.43 shows the magnitude-frequency relations of AE events located at $X = 80$–82 mm in Experiment B_1. The upper and the lower figures show the first half (No. 6-1–No. 6-13) and the second half (No. 6-14–No. 6-33), respectively. The left and the right figures show results in the loading and the unloading processes, respectively. V_p (dB) corresponds to the magnitude of an AE event. In this figure, it is remarked that the magnitude-frequency relation of AE events in the unloading process is significantly different from that in the loading process, that is, a large number of the largest class AE events occurred in the unloading process. Therefore, the magnitude-frequency relation for AE events in the unloading process cannot be expressed by the Ishimoto-Iida's equation.

Figure 7.44 shows the one-dimensional source locations of large AE events ($V_p \geq 55$ dB) along the X-axis in the unloading process, plotted as functions of the number of the analyzed AE data sets in Experiment B_1. This figure shows that the large events

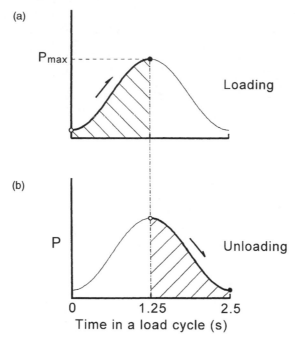

Figure 7.41.   Definition of (a) Loading and (b) Unloading process.

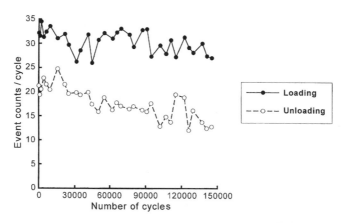

Figure 7.42.   AE events per one loading cycle as functions of the number of cycles in Experiment **B**₁. Solid and open circles correspond to loading and unloading process, respectively.

probably continued to occur at the same places. The waveforms and the amplitudes of these acoustic waves are quite similar, as shown in Fig. 7.45.

Next, the maximum stress level was elevated to 7 kN (Experiment **B**₂). In Fig. 7.46, the vertical axis is the time in a load cycle and the horizontal axis is the number of loading cycles. The AE events are plotted in this graph. The features of the distribution

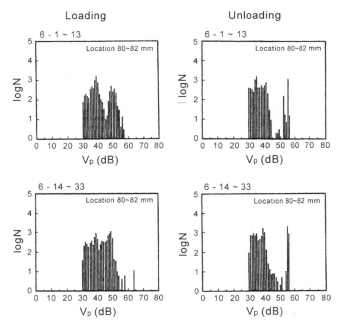

Figure 7.43. Magnitude-frequency relations of AE events located at X = 80–82 mm in Experiment **B**₁. The upper and the lower figures show the first half (No. 6-1–No. 6-13) and the second half (No. 6-14–No. 6-33), respectively. The left and the right figures show those in the loading and the unloading process, respectively. N indicates the number of AE events and V_p corresponds to magnitude.

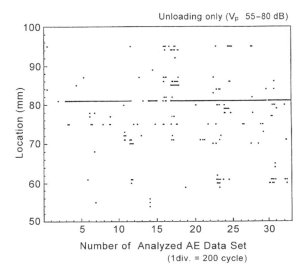

Figure 7.44. Source locations of large AE events ($V_p \geq 55$ dB) along the length of rock specimen in the unloading process are plotted as functions of the number of analyzed AE data set in Experiment **B**₁.

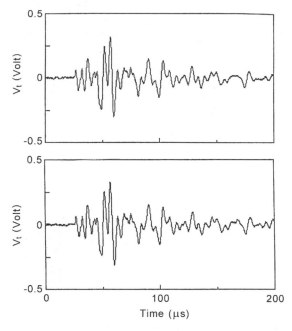

Figure 7.45.   An example of the waveform similarity of AE events under loading cycles.

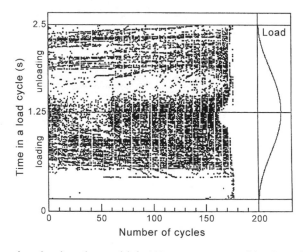

Figure 7.46.   Time in a load cycle at which AE events occurred is plotted as function of the number of cyclic loadings at the maximum stress level $P_{max} = 7\,kN$ (Experiment $\mathbf{B_2}$).

of AE events in this graph are quite different from those in Experiment $\mathbf{B_1}$, in the degree of AE activity and the pattern of AE event distributions.

Figures 7.47 (a) (b) show the AE event count rate as a function of the time during a loading cycle. Numerals on the right side indicate the data set numbers in Table 7.2. The top figure shows the temporal variation of an applied cyclic load. The most

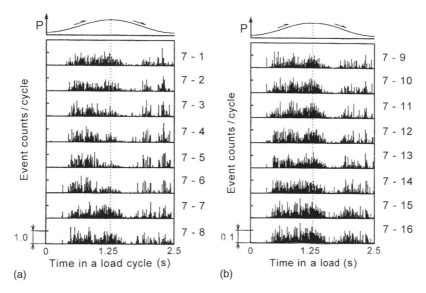

Figure 7.47. (a) AE event count rate as a function of time during a loading cycle. Top figure shows the temporal variation of load. Numerals on the right side indicate the data set numbers in Table 7.2. (b) Continued.

noticeable feature in this case is that a number of AE events occurred at the highest level of applied load and they increased with increasing number of loading cycles. As mentioned before, the AE activity was quite low around the maximum load in Experiment B_1.

Figures 7.48 (a) (b) show the AE activity indicated by $\Sigma \sqrt{E}$ (E: energy) as functions of the time during a load cycle. In this figure, the fine variations of the AE activity under cyclic loads, considering the energy released as acoustic waves, can be seen. Thus the process of concentration of AE activity around P_{max} was expressed clearly.

In Fig. 7.49, we defined three processes, namely (a) Loading process, (b) High-load process and (c) Unloading process, for the following discussion. Figure 7.50 shows the magnitude–frequency relation for AE events in the processes of Loading, High-load and Unloading in the left, center and right figures, respectively. Numerals in each figure (7-1, 7-6, 7-11, 7-16) indicate the data sets in Experiment B_2. N is the number of AE events and V_p corresponds to M (magnitude). Among these figures, the Ishimoto-Iida equation can be applied in the case of the High-load process, because log N decreases continuously with increasing V_p ($\sim$M). This result strongly suggests that AE events in the High-load process occurred by fracturing. On the other hand, in other two cases (Loading and Unloading processes), the Ishimoto-Iida equation cannot be applied, because the magnitude-frequency relation is discontinuous. This result suggests that AE events in these cases occurred by stick-slip in frictional sliding along the micro-crack surface, in addition of microfracturing.

Figure 7.51 shows the temporal variation of the AE count rate in the three processes, as functions of the number of loading cycles in Experiment B_2. In the cases of the

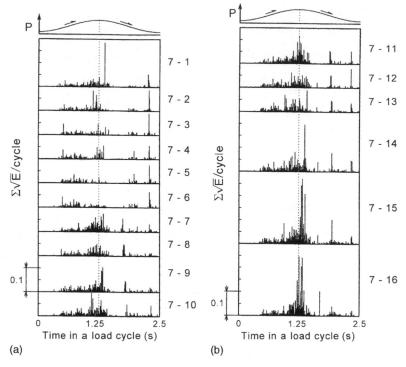

Figure 7.48.   (a) AE activity shown by $\Sigma\sqrt{E}$ as a function of time during a load cycle. Numerals on the right side indicate the data set numbers in Table 7.2. (b) Continued.

Loading and Unloading processes, the AE count rates are nearly similar and slightly increase with the increase of the number of loading cycles. However, the AE count rate in the High-load process markedly increased from around the 60th cyclic loading until the main fracture occurred.

Figure 7.52 shows $\Sigma\sqrt{E}$/cycle (E : energy released by AE) as functions of the number of loading cycles in Experiment B_2. In this figure, the marked increase of AE activity preceding the main fracture can be seen quite clearly. These curves for the Loading and Unloading process do not change significantly.

Figure 7.53 shows one-dimensional source locations of large AE events ($V_p \geq 63$ (dB)) as a function of the number of loading cycles. The number of large AE events began to increase from about 70th cycle and AE sources concentrated in the region of X = 70 mm and its surrounding area.

I want to comment about the source locations in this experiment. Just before the main fracture, AE sources did not concentrate in a limited region. This scatter of AE sources may be attributed to the shape of the tested specimen to some degree, that is, the thickness or height (40 mm) was relatively large in comparison with the width (30 mm) and the inner span (40 mm) (see Fig. 7.32). In the case of the preceding section 4.2, the thickness (30 mm) was much smaller than the width (50 mm) and the inner span (60 mm).

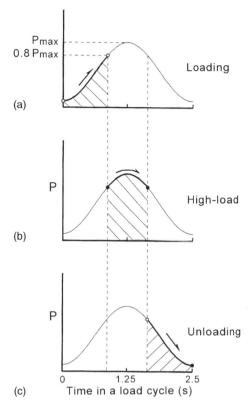

Figure 7.49.   Definition of (a) Loading process, (b) High-load process and (c) Unloading process.

7.3.d   *Concluding remarks*

The cyclic bending experiments of a rock were carried out with loading of a tensile type. The tested granodiorite is a typical hard crystalline rock that includes various kinds of minerals. Since the strength of these minerals is different, the rock has a non-uniform structure. Therefore, the specimen include many weak points from which new cracks may start and then develop. The observed results show that the generation of AE events in the initial stage of cyclic loading can be explained by this mechanism. These AE events occurred at each load level, that is, they are a load level dependent type (Type 1).

Under successive application of the cyclic loading of the same level, AE events also continued to occur at the same place. The mechanism of AE generation of this kind may be attributed to stick-slip in frictional sliding along new or pre-existing crack surfaces, as mentioned above (Type 2).

The occurrence of AE events at about the peak load in Experiment B_2 is somewhat different from Type 1, because these AE events occurred in a macroscopic fracturing

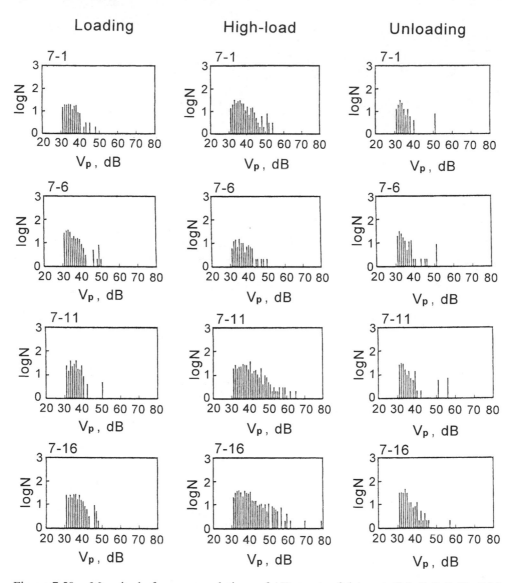

Figure 7.50.   Magnitude-frequency relations of AE events of data sets 7-1, 7-6, 7-11, 7-16 in Experiment **B**$_2$. The left, the center and the right figures show those in the loading, the high-load and the unloading processes, defined in Fig. 7.49. N indicates the number of AE events and V$_p$ corresponds to magnitude.

process and they sometimes are related to the following major fracture. They are classified as Type 3.

Although I discussed the results of a great number of experiments under cyclic loading conditions in some detail, I would like to emphasize again the importance of the following findings mentioned in Chapter 4, AE events continue to occur under

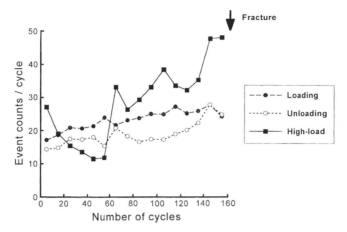

Figure 7.51. AE event count rate as a function of the number of loading cycles in Experiment B_2. The temporal variations of AE activity of the three different processes (Loading, Unloading and High-load) and the main fracture are shown.

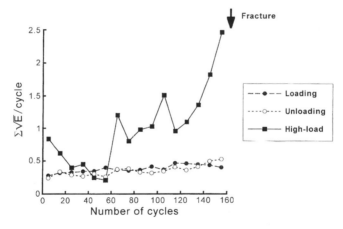

Figure 7.52. Energy released by AE as a function of the number of loading cycles in Experiment B_2.

a constant load, and sometimes a major fracture occur. This phenomenon is called "static fatigue". In Experiment B_1, a great number of cyclic loadings (145,000 cycles) continued to be applied to the rock specimen. In this experiment the AE count rate did not show any change during this long period of cyclic loads. However, we examined the amplitudes of acoustic waves (the energy released by acoustic emission) and found that the released energy began to change gradually and markedly from the 44,550th cycle. This unexpected result suggests that important information about the fine variation of mechanical properties of rocks and other materials may be obtained by use of the AE technique.

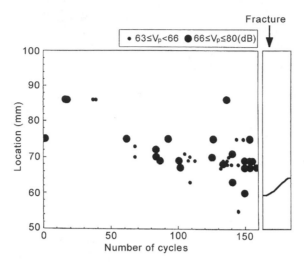

Figure 7.53.   Source locations of large AE events as a function of number of loading cycles in Experiment **B**$_2$. The main crack is shown on the right side.

REFERENCES

Goodman, R. E. (1963). Subaudible noise during compression of rocks, Geological Society of America Bulletin, **74**, 487–490.

Ishibashi, K. (1980). Observation of the 1980 Izu-Hanto-Toho-Oki earthquake swarm, (1) Epicentral distribution off aftershocks just after the earthquake (M6.7) at 16 h 20 m June 29. Abstr. Seismol. Soc. Japan, 1980, No. 2, p.6. (in Japanese).

Japan Meteorological Agency. (1980). Earthquake swarm off the east coast of central Izu Peninsula, June–July, 1980. Rep. Coord. Comm. Earthquake Prediction, **25**, 134–140. (in Japanese).

Kaiser, J. (1953) Erkenntnisse und Folgerungen aus der Messung von Gerauschen bei Zugbeanspruchung von Metallischen Werkstoffen, Archiv für das Eisenhüttenwesen, **24**, 43–45.

Karakama, I., I. Ogino, K. Tsumura, K. Kanjo, M. Takahashi and R. Segawa. (1980). The earthquake swarm east off the Izu Peninsula of 1980. Bull. Earthq. Res. Inst., Tokyo Univ., **55**, 913–948. (in Japanese).

Matsu'ura, R.S. (1983). Detailed study of the earthquake sequence in 1980 off the east coast of the Izu Peninsula, Japan. J. Phys. Earth, **31**, 65–101.

Mogi, K. (1966). Some precise measurements of fracture strength of rocks under uniform compressive stress. Rock Mech. Eng. Geol., **4**, 43–55.

Mogi, K. (1983). Relation between the 1980 Izu-Hanto-Toho-Oki earthquake (M6.7) and earthquake swarms before and after. Abstract Seismol. Soc. Japan, 1983, No. 1, p.23. (in Japanese).

Mogi, K. (1985). Earthquake Prediction. Academic Press, Tokyo, pp.355.

Mogi, K. and H. Mochizuki. (1982). Observation of high-frequency seismic waves by a hydrophone directly above the focal region of the 1980 earthquake (M6.7) off the east coast of the Izu Peninsula, Japan. Earthq. Prediction Res., **2**, 127–148.

Mogi, K. and Y. Mori. (1993). Experimental study of AE in rock specimens under cyclic loadings. Research Progress Report of a Grant-in Aid for Scientific Research from the Ministry of Education, Science and Culture, p. 87. (in Japanese).

Mori, Y., K. Saruhashi and K. Mogi. (1994). Acoustic emission in rock specimen under cyclic loading. Progress in Acoustic Emission VII, Japanese Society for Non-Destructive Inspection, 173–178.

Ohtake, M., M. Imoto, M. Ishida, T. Okubo, Y. Okada, K. Kasahara, M. Tachikawa, S. Matsumura, S. Yamamizu, and K. Hamada. (1980). Izu–Hanto-toho-oki earthquake of 29 June 1980 and seismic activity before and after it. Abstr. Seismol. Soc. Japan, 1983, No. 2, p. 4. (in Japanese).

Saruhashi, K. (1994). Study of AE in rock specimens under cyclic loadings. Thesis for Master Degree from Nihon Univ., pp. 81. (in Japanese).

Saruhashi, K., L.W. Zheng, Y. Mori and K. Mogi. (1993). AE in rock under cyclic loading. Proc. 1993 National Conf. Acoustic Emission, Japanese Soc. Non-Destructive Inspection, 109–114. (in Japanese).

Yoshikawa, S. and K. Mogi. (1981). A new method for estimating the crustal stress from cored rock samples: laboratory study in the case of uniaxial compression. Tectonophysics, **74**, 323–339.

Yoshikawa, S. and K. Mogi. (1990). Experimental studies on the effect of stress history on acoustic emission activity – A possibility for estimation of rock stress. J. Acoustic Emission, **8**, 113–123.

Zienkiewicz, O. C. (1971). The Finite Element Method in Engineering Science. McGraw-Hill, New York.

III

Rock Friction and Earthquakes

CHAPTER 8

Laboratory experiment of rock friction

8.1   INTRODUCTION

The study of friction is very important in rock mechanics. Friction plays an important role in the sliding process along the discontinuous plane at various scales. According to Soda (1971), the friction phenomena have been studied as the resistances to sliding of solid materials by many investigators, particularly Leonardo da Vinci (1452–1519), Amonton (1663–1705), and Coulomb (1736–1806). The modern knowledge of frictional phenomena is mainly attributed to the work of Bowden and Tabor (1950, 1964).

Coulomb established the following friction law by his very careful experiments:

(1) The frictional force is proportional to the normal load, and it is independent of the size of the surface in contact.
(2) The kinetic frictional force is independent of the sliding velocity.
(3) The static frictional force is larger than the kinetic frictional force.

The Coulomb law is sometimes called "Amonton's law" or "Amonton – Coulomb law". The law is still fundamental in friction studies.

The research on friction and sliding phenomena in rocks are well reviewed in Jaeger and Cook (1976), Paterson (1978), and Scholz (1990). Figure 8.1 shows common methods in friction studies. There are a variety of different methods which have advantages and disadvantages. A conventional direct shear test (a) is simple and suitable for large specimens, and it has been used widely, for example, by Coulomb. However, this system contains a moment, and so the normal stress is not uniform over the surface. The conventional double-shear test (b) also contains a moment, and so the normal stress is not uniform on the sliding plane, and it has two sliding surfaces. In the biaxial compression method shown in (c), uniform normal stresses can be applied, but frictional sliding experiments under controllable non-uniform normal stresses are impossible and constant normal stresses cannot be kept for stick-slip.

The conventional triaxial compression test of a cylindrical saw-cut rock sample (d) is widely used, particularly for friction experiments under high normal stress conditions. But it has the disadvantage that this configuration is unsuitable for the case of a large amount of slip and for dynamic measurements related to stick-slip.

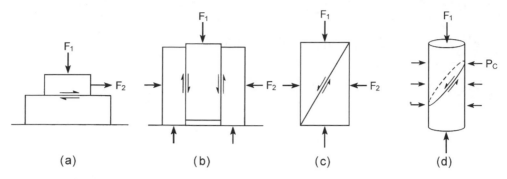

Figure 8.1(a)–(d).   Common methods of friction tests : (a) conventional shear test; (b) conventional double-shear test; (c) sliding on saw cut in biaxial test; (d) sliding on saw cut in conventional triaxial test.

Another important method is the type using a rotary shear apparatus by which the effects of very large amounts of sliding on the surface are studied.

Over two hundreds years ago, Coulomb obtained the following equation (e.g. Soda, 1971):

$$F = A + \mu_1 P \qquad (8.1)$$

where F is the frictional force, P is the normal force, and A and μ_1 are constants. This simple equation is still used at present (e.g. Jaeger, 1959; Byerlee, 1978). Byerlee (1978) obtained the following relation on the basis of his many experimental data on rock friction:

$$\tau = 50 + 0.6\sigma_n \qquad (8.2a)$$

for $\sigma_n > 200\,\mathrm{MPa}$, and

$$\tau = 0.85\sigma_n \qquad (8.2b)$$

at lower normal stress. This friction law can be applied to various kinds of rocks, so it is called Byerlee's law.

8.2   NEW DESIGN OF A DOUBLE-SHEAR TYPE APPARATUS

Figure 8.2 shows advantages and disadvantages in friction tests (b) and (c) shown in Fig. 8.1. In the conventional double shear method on the left, a nearly constant normal stress is maintained, but the shear stress is not uniform along the fault (see Fig. 8.7). In the shear experiment by biaxial compression, on the right, a uniform shear

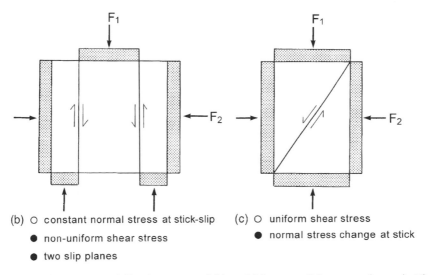

(b) ○ constant normal stress at stick-slip
    ● non-uniform shear stress
    ● two slip planes

(c) ○ uniform shear stress
    ● normal stress change at stick

Figure 8.2.   Advantages and disadvantages of (b) and (c) types of shear test shown in Fig. 8.1.

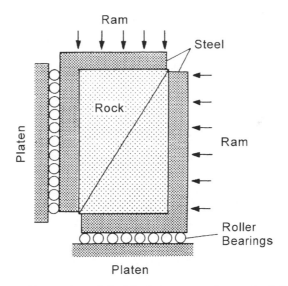

Figure 8.3.   Schematic illustration of the biaxial compression method (Scholz et al., 1972). Uniform shear stress is applied to the diagonal saw cut plane.

stress can be applied, but the normal stress changes abruptly during a sudden frictional slip in the stick-slip process, and the application of a controllable nonuniform normal stress is impossible. The application of the controllable nonuniform normal stress is important in the study related to the behavior of asperities in earthquake sources.

For fundamental studies of precursory phenomena of earthquakes, we measured some characteristics of precursory strain changes prior to sudden frictional slip along

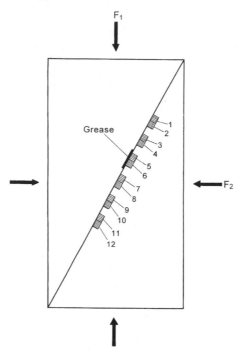

Figure 8.4.   Strain gages (Nos. 1–12) attached to the sample in order to measure deformation in the vicinity of the saw cut plane. Grease is applied in some cases in order to reduce locally the frictional strength of the sliding plane. (Mogi, 1981).

an artificial fault. Figure 8.3 shows schematically the biaxial compression method by Scholz et al. (1972), where uniform shear stress is applied to the diagonal saw cut plane. I conducted an experiment using their machine (Mogi, 1981). Figure 8.4 shows the sample and strain gages used in the experiment. When lateral pressure is held constant and force in the vertical direction is increased, a sudden slip following a stick (a stick-slip event) occurs along the fault. This sudden slip is thought to correspond to seismic fault movements along an existing fault. Figure 8.5 shows the temporal changes in strain parallel to the fault in the case of a homogeneous fault. Strains increased linearly, but no noticeable anomalies were observed prior to a sudden slip for this sensitivity in strain measurements. Figure 8.4 shows the case of a heterogeneous fault. In this case I applied a thin coating of grease around gage no.6 and partially reduced the frictional strength. Figure 8.6 shows the temporal changes in strain in a direction parallel to the fault trace at each place. Differing from the homogeneous fault, obvious anomalies in the deformation curve appear preceding the slip. This strain anomaly is most striking at gage no. 6, becomes less obvious with distance, and no anomaly can be observed at gages no.2 and 12. This distribution makes it clear that the precursory deformation is due to the mechanical heterogeneity of the fault plane.

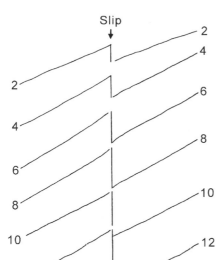

Figure 8.5. Temporal changes in strain parallel to the fault in the case of a homogeneous fault. The numerals correspond to the numbers of the strain gages in Fig. 8.4. (Mogi, 1981).

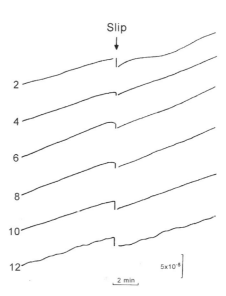

Figure 8.6. Temporal changes in strain parallel to the fault in the case of a heterogeneous fault. (Mogi, 1981).

To study various features of frictional sliding in fault of rocks, we designed a new double-shear type loading machine (Mogi and Mochizuki, 1990).

The top figures in Fig. 8.7 show schematically the case of the conventional double-shear type loading machine, which was explained above. This is a symmetrical system in which one block is pushed by F_1 between two other blocks which are pushed by the lateral force (F_2), as shown by the arrows. In this system, the stress distribution along the fault is complex, as shown simply by broken curves in the top figure, and so the shear stress by the increase of the axial force F_1 is not uniform. In the right figure, strains along the fault are shown as a function of time (numerals correspond to strain gages along the fault).

The bottom figures show schematically the case of the new system. The two important revisions are as follows:

(1) In the conventional method, the axial force was applied only at the top surface of the central block through a steel end piece, so the axial stress in the central block markedly decreased in the lower portion. In the new apparatus, the axial force (F_1) is not directly applied to the top surface of the central block, but applied through the steel blocks, as explained below. By this method, the central block moves uniformly as a whole, and a uniform shear stress along the fault planes can be applied, as shown in the right figure.

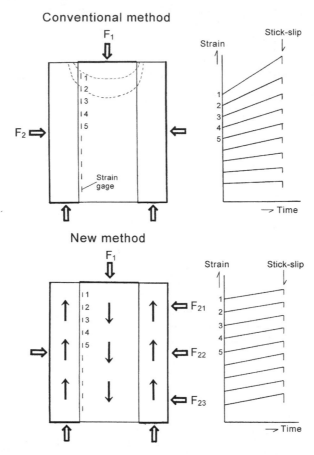

Figure 8.7. Comparison between the new double-shear type loading method and the conventional method shown in Fig. 8.1 (b).

(2) On the other hand, frictional experiments in which the normal stress is not uniform, are also important, as mentioned above. In the new apparatus, different normal stresses can be applied by three independently controlled jacks (F_{21}, F_{22}, F_{23}). Thus, systematic friction experiments about a controllable heterogeneous fault are possible. Such control of normal stress in friction experiments is impossible in other methods shown in Fig. 8.1.

For continuation of a large amount of sliding and many stick-slip experiments under nearly the same stress conditions, we designed a relatively large-scale apparatus in which a large rock sample with a fault length of 1 meter can be used.

Figure 8.8 shows of the rock sample and the loading system. A rectangular parallelepiped granite sample with 1 m length and 10 cm thickness was cut into three blocks and joined together again. The central block is sandwiched between the two stiff steel blocks and they are strongly bonded by adhesive and vertical steel bolts.

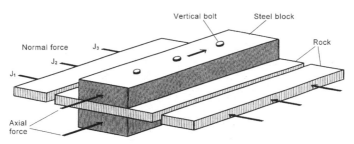

Figure 8.8.   Schematic view of rock sample and loading system in the new double-shear type testing machine for friction experiment. The central block of rock sample is sandwiched between two steel blocks, and they are strongly bonded by adhesive and vertical steel bolts. Axial force is applied to the steel blocks. Normal force is applied by three jacks (J_1, J_2 and J_3) which can be controlled independently, and so both uniform and nonuniform shear stresses can be applied. Length and thickness of rock specimens are 100 cm and 10 cm, respectively.

These rock blocks are compressed by a biaxial compression apparatus. In this case, the axial force is applied to the stiff steel blocks directly. Thus the axial force in the central rock block is applied uniformly through these steel blocks. The normal force is applied by the three jacks (J_1, J_2 and J_3) which can be controlled independently, and so both uniform and nonuniform shear stresses can be applied.

Figure 8.9 shows a horizontal view of this double-shear type loading apparatus and the rock specimen used for friction measurements. The axial force is applied by a 3000 kN jack. The normal force is applied by three 100 kN jacks, as mentioned above. Strains at many points along the fault are measured and acoustic signals are measured as well. Figure 8.10 shows the double-shear type testing machine. The right figure shows the apparatus for biaxial compression and the left-bottom figure shows a rock specimen sandwiched between steel blocks and other devices for shear experiments. Figure 8.11 shows a view of the main part of the friction experiment.

In conclusion, the advantages of this method are as follows:

(1) A nearly uniform shear stress along the fault can be applied.
(2) Since the length of the fault is relatively large, the effect of large amount of sliding can be investigated under a constant stress state.
(3) Since the normal stress does not change during stick-slip, accurate measurements of slip under a constant normal stress are possible and many slip experiments can be repeated under a constant stress state.
(4) Since the normal force is applied by the three jacks independently, a controllable nonuniform normal stresses can be applied to the fault plane.

It should be noted that the double-shear type test has two sliding planes. However, this is not a serious disadvantage. If a suitable ductile material, such as Teflon, is inserted into one fault plane, the frictional sliding and stick-slip occur only along a single fault, as shown by Yoshida and Kato (2001). It is quite favorable that this machine is a symmetric and stable system.

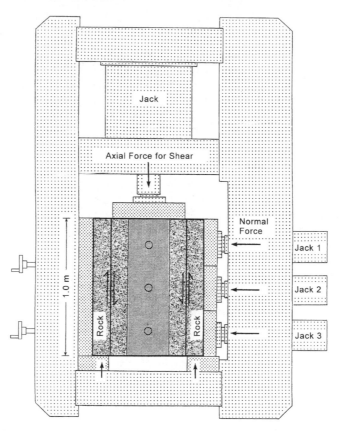

Figure 8.9.   The horizontal view of the new double-shear type testing apparatus. The length of fault planes is 1.0 m. Normal force is applied by J_1, J_2 and J_3.

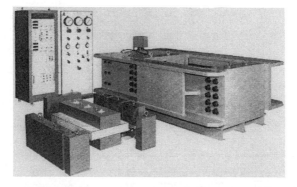

Figure 8.10.   A set of the double-shear type testing machine. Right: biaxial compression apparatus; left-bottom: rock specimen sandwiched between steel blocks and other devices for experiments.

Figure 8.11.   Friction experiment using the new double-shear type machine.

8.3   EXPERIMENTAL RESULT

In this experiment, Inada granite was used. The fault surface was not polished very well, so it was rather a heterogeneous fault, as will be shown below.

Figure 8.12 shows the relation between the axial force and time under a constant displacement rate of the axial piston. Stick-slip events occurred successively on the sliding surface. In the initial stage, drops of axial force were not constant, but after the repetition of stick-slip events the force-time curve becomes very regular. Figure 8.13 shows the temporal variations of the maximum and minimum values of the axial force. After some number of stick-slip events, these values become nearly constant. Such stick-slip phenomena have been observed in many materials. Since about 1966, these phenomena in rocks have attracted particular attention in relation to the earthquake mechanism. Brace and Byerlee (1966) proposed stick-slip as the mechanism of earthquakes. According to their opinion, because earthquakes are recurring slip instabilities on preexisting faults, earthquakes correspond to stick-slip phenomena. I will discuss this issue later. Thus, stick-slip phenomena have been studied by a number of researchers (e.g. Byerlee and Brace, 1968; Dieterich, 1972; Scholz, Molnar and Johnson, 1972).

In this section, the measurements of precursory strain changes prior to sudden slips along the fault using the above-mentioned shear apparatus is presented. The left figure in Fig. 8.14 shows the temporal variation in linear strains parallel to the fault. The right figure shows the distribution of the normal stress obtained by the following press-film method. The color of the press-film (originally white) changes to red and

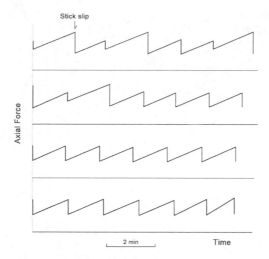

Figure 8.12.   Force – time curve when the axial force is applied nearly at a constant rate. Stick-slips occurred successively along the sliding plane.

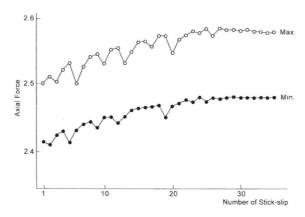

Figure 8.13.   Temporal variations of the maximum value and minimum value of the axial force in the process of successive stick-slips.

the intensity of the red color is proportional to the degree of compression. There are three films of different sensitivities, so this is very useful for rough measurements of stress-distribution. The right figure shows that this fault plane is not completely flat and it contains two strongly compressed regions, that is *asperities*. Numerals in this figure (1–16) indicate the positions of strain gages along the fault. The left figure shows that marked precursory strain curves prior to the sudden slip were observed in two groups of strain gages, No.3, No.4, No.5 and No.13, No.14, No.15. These two groups of anomalous strain curves are very similar. A similar pattern was reported in a previous paper (Mogi et al., 1982), in which experiments was conducted by the

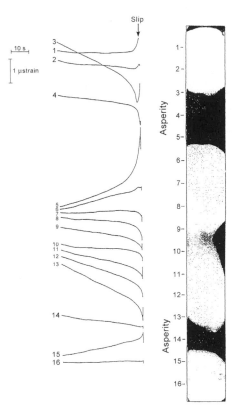

Figure 8.14. Temporal variation of strain parallel to the heterogeneous artificial fault in the case of high loading rate. The right figure shows the distribution of the normal stress in the fault plane. Black regions are highly compressed regions (asperities).

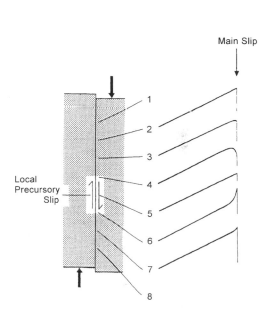

Figure 8.15. Typical strain curves prior to a sudden main slip of the fault which is attributed to local precursory slip. (Mogi et al., 1982)

conventional double-shear type method. In the paper, we explained that this pattern occurred by local precursory slip. The typical pattern is represented in Fig. 8.15. The anomalous strain curves in Fig. 8.14 are quite similar to those in Fig. 8.15 and so these anomalous patterns in Fig. 8.14 can be attributed to the occurrence of pre-slip at asperities prior to the main slip.

Figure 8.16 shows the temporal variation in linear strains along the same fault, in which the axial loading rate is significantly low. In this figure, strain curves of the whole process before and after the main slip are shown. The precursory strain changes related to the two asperities are similar to those shown in Fig. 8.14.

Figure 8.17 shows another case. The right figure shows the distribution of normal stress, which is different from that in Figs. 8.14 and 8.16. The left figure shows the temporal variation of linear strain parallel to the fault before and after a main sudden slip, in the case of a high loading rate. In this case, the highly compressed region

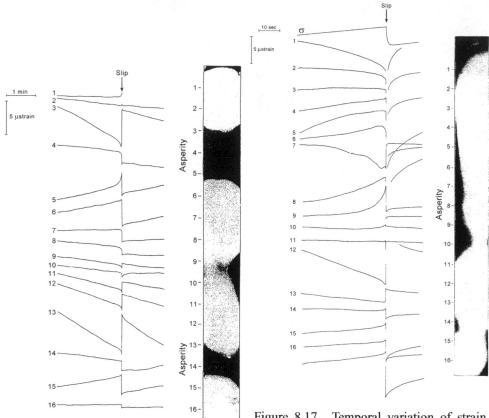

Figure 8.16.   The same as Fig. 8.14, but the loading rate is significantly lower.

Figure 8.17.   Temporal variation of strain parallel to the fault before and after a main sudden slip. The distribution of normal stress is different from the case in Fig. 8.14.

(asperity) locates in the central region (strain gage No. 6–No.11), and remarkable precursory strain changes were also observed in the central region (No.6–No.9).

Figure 8.18 (left) shows the temporal variation of linear strain along the same fault, but the loading rate is lower than in the case shown in Fig. 8.16. The right figure shows only the very high normal stress region (asperity) in the fault plane, which was observed by the press-film method with low sensitivity. The precursory change is significantly lower than that in the case in Fig. 8.17, but it was also observed by strain gages No.6, No.7 and No.8 located in the asperity region.

The above-mentioned results suggest that the main slip of the fault may be closely related with the pre-slip in the asperity region.

In Fig. 8.19, the magnitude of precursory strain prior to sudden slips is plotted against the number of the stick-slip events. The magnitude gradually decreases with the repetition of stick-slip. The axial loading was stopped after N = 58 and the reloading started after a 4 day interval. The precursory strain clearly increases after the

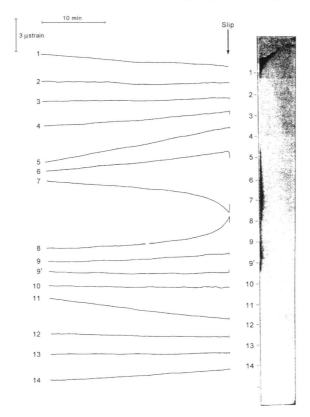

Figure 8.18.   The same as Fig. 8.17, but the loading rate is very low. The right figure shows only the distribution of regions where the normal stress is very high.

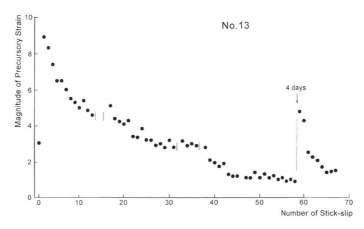

Figure 8.19.   Magnitude of precursory strain is plotted against the number of stick-slips. After the 4 days interval, the precursory strain increased.

time interval and then rapidly decreases. This result suggests that some mechanical changes occur on the compressed fault plane with time. About this problem, further investigations are needed.

[As mentioned before, I was appointed as the director of Earthquake Research Institute, University of Tokyo and then moved to Nihon University, and so I could not continue our laboratory experiments. Fortunately, Yoshida and his colleague (e.g. Yoshida and Kato, 2001) are conducting laboratory experiments using this machine.]

REFERENCES

Bowden, F. P. and D. Tabor. (1950). The Friction and Lubrication of Solids. Vol. I, Clarendon Press, Oxford.

Bowden, F. P. and D. Tabor. (1964). The Friction and Lubrication of Solids. Vol. II, Clarendon Press, Oxford.

Brace, W. F. and J. D. Byerlee. (1966). Stick-slip as a mechanism for earthquakes. Science, 153, 990–992.

Byerlee, J. D. and W. F. Brace. (1968). Stick-slip, stable sliding, and earthquakes – effect of rock type, pressure, strain rate, and stiffness. J. Geophys. Res., 73, 6031–6037.

Dieterich, J. H. (1972). Time dependent friction in rock. J. Geophys. Res., 77, 3690–3697.

Jaeger, J. C. and N. G. W. Cook. (1976). Fundamentals of Rock Mechanics. Chapman and Hall, London, pp.585.

Mogi, K. (1981). Earthquake prediction and rock mechanics. J. Soc. Material Sci. Japan, 30, 105–118. (in Japanese).

Mogi, K., S. Yoshikawa and S. Ogita. (1982). Changes of the sudden slip of artificial fault planes (2) – precursory deformation. Abstr. Seismol. Soc. Japan, No.1, p.128. (in Japanese).

Mogi, K. and H. Mochizuki. (1990). Measurements of precursory strain prior to sudden slips along a heterogeneous artificial fault using a new double-shear type loading machine. Int. Symp. on Earthquake Source Physics and Earthquake Precursors, (Extended Abstr.), 43–47.

Paterson, M. S. (1978). Experimental Rock Deformation – The Brittle Field. Springer-Verlag, Berlin, pp.254.

Scholz, C., P. Molnar and T. Johnson. (1972). Detailed studies of frictional sliding of granite and implications for earthquake mechanism, J. Geophys. Res., 77, 6392–6406.

Scholz, C. (1990). The Mechanics of Earthquakes and Faulting. Cambridge Univ. Press, pp.439.

Soda, N. (1977). Stories of Friction. Iwanami-shoten, pp.214. (in Japanese).

Yoshida, S. and A. Kato. (2001). Single and double asperity failures in a large-scale biaxial experiment. Geophys. Res. Lett., 28, No. 3, 451–454.

CHAPTER 9

Typical stick-slip events in nature and earthquakes

9.1 INTRODUCTION

The author's academic research started in 1954 as a Research Associate in the laboratory of physical volcanology of the Earthquake Research Institute, University of Tokyo. In the beginning in 1957, I published a paper titled "On the relation between the eruption of Sakurajima Volcano and the crustal movements in its neighborhood" (Mogi, 1957, in Japanese), in which I proposed a model to explain the crustal deformation caused by a volcanic eruption. However, there was no response to the paper in Japan. In the next year, I wrote a similar, but more extended paper (Mogi, 1958) in English. Immediately a number of volcanologists in the world had interest and they applied this model to similar problems at many volcanoes, which is continuing to the present. Ryan et al (1983) wrote "Mechanical models of the internal structure of active volcanic systems dates from the work of Mogi (1958)" at the beginning of the historical review in their paper. This model is quite simple and applicable to many volcanoes, and it is often called the *Mogi model.*

Below, I will discuss in some detail about this model and its background.

As shown in Fig. 9.1, Sakurajima Volcano which is an andesitic volcano, is located in Kagoshima Bay, southern Kyūshū, Japan. The 1914 Sakurajima Volcanic eruption was the largest eruption in the past 100 years in Japan. Two branches of lava flowed to the eastern and the western coasts of the Sakurajima island and covered an area of 24 km^2, and the total volume of the ejected materials amounted to 2 km^3. On the morning of 11 January 1914, a number of earthquakes, including a M 7.0, began to occur and then the great eruption occurred.

After the eruption, the marked crustal deformation around the volcano was found by leveling and triangulation surveys by the Geographical Survey Institute. These results of the crustal deformation related to the volcanic eruption were reported in detail by Ōmori (1914–1922). According to the report, a very wide area of southern Kyūshū subsided. Particularly, it was noted that the contours of equal depression were circular and its center located in Kagoshima Bay, which corresponds to the Aira caldera, 10 km north of the crater of Sakurajima Volcano (Fig. 9.2). When I read Ōmori's paper in 1956, I perceived immediately that the magma reservoir may be located below the center of the depression, and it may be caused by the decrease of pressure of the magma chamber due to the outflow of an enormous volume of lava from Sakurajima Volcano on the rim of Aira caldera. I thought that this idea is quite simple and so my idea may not be a new one. However, this kind of the explanation

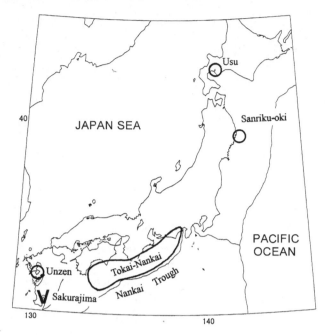

Figure 9.1.   Locations of Sakurajima Volcano and the four regions (Usu Volcano, Unzen Volcano, Sanriku-oki region and Tokai-Nankai region) in Japan, mentioned in this chapter. In the four regions, typical characteristic earthquakes occurred repeatedly by stick-slip mechanism.

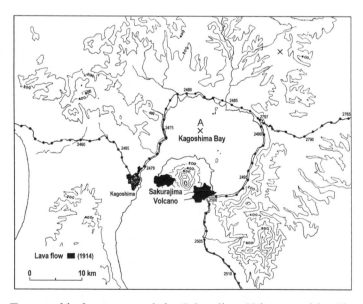

Figure 9.2.   Topographical map around the Sakurajima Volcano and leveling route in the area. Black area: lava flow in the 1914 great eruption of Sakurajima Volcano. A: center of depression.

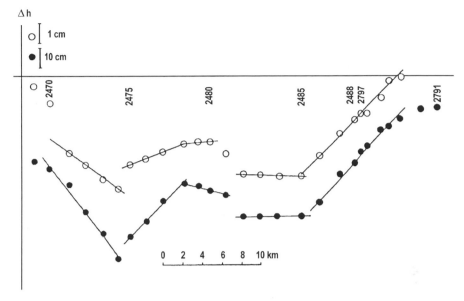

Figure 9.3. The vertical displacements of bench marks along the leveling route in the Kagoshima Bay area. Solid circle: 1895–1914; open circle: 1914–1915 (from Tsuboi, 1929). Tsuboi tried to explain the depression in this region by "Block movement hypothesis".

about the great depression related to the 1914 eruption and the quantitative analysis of the depression from a volcanological standpoint had not been presented before.

Since the crustal movement related to the 1914 Sakurajima eruption was a very important event, experts on deformation problems published many research papers about the observed results. In particular, Tsuboi (1929) and Miyabe (1934) discussed this event from the viewpoint of the block movement hypothesis. Figure 9.3 represents the typical result reported by Tsuboi (1929). The vertical displacements of bench marks are plotted along the leveling route in the Kagoshima Bay area. He thought that the discontinuous points in the vertical movement along the leveling route might be the boundary of each block, and he argued that the effect of loading by a mass of lava might also yield the general depression by tilting of blocks. After that, a number of researchers followed Tsuboi's idea, and this explanation continued until 1956.

Thus, I proposed the above-mentioned model (inflation–deflation model) using the elastic calculation by Yamakawa (1955), developed in relation to the earthquake source mechanism. According to elasticity theory, the calculated deformation of the surface of a semi-infinite elastic body, caused by the change of hydrostatic pressure in a small sphere at depth (f) is as follows, (see Fig. 9.4)

$$\Delta d = \frac{3a^3 p}{4\mu} \frac{d}{(f^2 + d^2)^{3/2}} \tag{9.1}$$

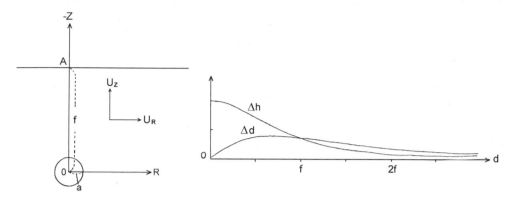

Figure 9.4. Calculated deformation of the semi-infinite elastic body which is caused by the change of the hydrostatic pressure in a small sphere in the semi-infinite elastic solid (Yamakawa, 1955).

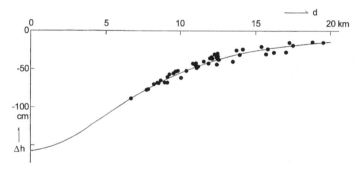

Figure 9.5. Relation between the observed vertical displacements (Δh) of bench marks and the distances (d) from the center of depression (A) to bench marks. The curve is calculated from the best-fit model with a spherical pressure source at a depth of 10 km.

$$\Delta h = \frac{3a^3 p}{4\mu} \frac{f}{(f^2 + d^2)^{3/2}} \qquad (9.2)$$

in which   d: radial distance on the surface from the point A.

            f: depth of the center of the sphere.

            P: change of the hydrostatic pressure of the sphere.

    $\mu(=\lambda)$: Lame's constant

        Δ d: displacement in the direction of the R-axis on the surface.

        Δ h: vertical displacement on the surface.

    Thus, if the radius (a) of a pressure source is relatively small, the form of deformation curves is a function of depth and values of displacements depend on P, a and μ.

    Figure 9.5 shows the relation between the vertical displacements of bench marks and the distances (d) from the center of depression (A) to the bench marks. The observed data (solid circles) and the best-fit calculated curve with a pressure source

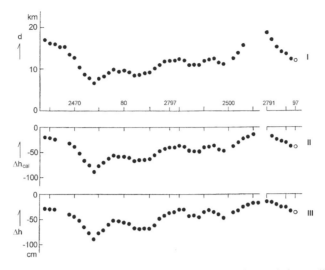

Figure 9.6. Relation between the vertical displacement (Δh) and the radial distance (d). (I): radial distance (d) of bench mark from the center of depression. (II): calculated vertical displacement (Δh_{cal}) at the station of the radial distance (d). (III): observed vertical displacement of bench mark along the leveling route.

at a depth of 10 km are shown. Figure 9.6 also shows a good agreement of the observed data and the calculated values by this model. The horizontal axis is the number of the bench mark along the leveling route in Kagoshima Bay region shown in Fig. 9.2. d is the radial distance of the bench mark from the center of depression. Δh is the observed vertical displacement and Δh_{cal} is the vertical displacement calculated using the elastic model with a pressure source at a depth of 10 km. We cannot find any large difference between Δh and Δh_{cal}. This result shows that the great depression can be explained by an elastic model with a pressure source at depth of 10 km and that the "Block Movement" proposed by Tsuboi and others was not necessary. The apparent boundaries between "blocks" in Fig. 9.3 correspond only to the points at which the directions of the leveling routes abruptly change.

From the ground deformation mentioned above, I proposed the following process of the 1914 great eruption of Sakurajima Volcano. The profile of the Sakurajima district is shown in Fig. 9.7. As mentioned above, Sakurajima Volcano is situated on the rim of Aira Caldera and the magma reservoir (its center) is at a depth of 10 km under Kagoshima Bay. In the period preceding the 1914 eruption, the pressure in the magma reservoir continued to increase and the inflation of the ground continued. In 1914, an enormous volume of magma extruded along the weak wall related to the caldera fractures (earthquakes) and lava flowed out from Sakurajima Volcano. As a result of out-flow of lava, the pressure in the magma reservoir decreased and a wide area significantly subsided (deflation of the ground).

Figure 9.8 shows the relation between the volcanic activity of Sakurajima Volcano and the volume (V) of inflation or deflation of the ground which is dependent on the

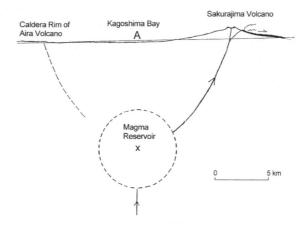

Figure 9.7.   Magma reservoir of Sakurajima Volcano and the mechanism of the eruption of the volcano deduced from the ground deformation.

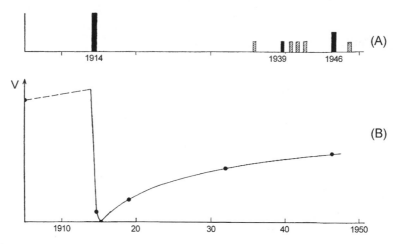

Figure 9.8.   Relation between the volcanic activities of the Sakurajima Volcano and the inflation /deflation in the Kagoshima Bay region. V is volume of the inflation (or deflation) in this region estimated by the leveling survey.

pressure of the magma reservoir, estimated from the data of leveling surveys. In this figure, it can be recognized that the surface volcanic activity is clearly affected by the increase of pressure in the magma reservoir. This result suggests that crustal deformations may provide clues for the prediction of volcanic eruptions. This model has been applied successfully to a number of volcanoes in the world. Figure 9.9 shows a recent result obtained by radar interferometry (Amelung, Jonsson, Zebker and Segall, 2000).

When I began this work in volcanology, I looked at a great number of characteristic seismogram in our laboratory, and I was strongly impressed by the high regularity in these seismograms. These were obtained by Prof. Minakami and his collaborators. They conducted seismic observations during the 1944 eruption of Usu Volcano,

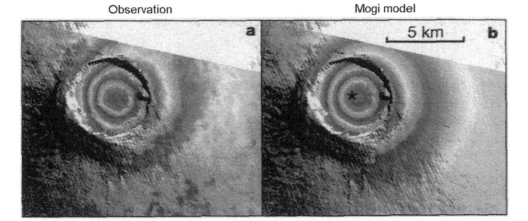

Figure 9.9.   Deformation at Darwin, Galapagos Islands, during 1992–1998. (a) Observation by radar interferometry. (b) Predicted interferogram from the best-fit Mogi model with source location in the center of the caldera (star) and depth of 3 km (Amelung et al., 2000). This is reproduced from the original color figure, Fig. 3 in Nature (vol 407, p. 995, 2000).

Hokkaido, Japan. I think that these seismograms may be the first records of characteristic earthquakes due to stick-slip events in the natural environment. I will further explain this in the following section.

9.2   USU VOLCANO AND UNZEN VOLCANO, JAPAN

Usu Volcano is located in southwestern Hokkaido, as shown in Fig. 9.1. The volcano is one of the most active volcanoes in Japan, and seven major eruptions have been recorded since 1663. At the end of 1943, felt earthquakes began to occur and this activity was followed by uplift of a wheatfield at the eastern foot of the volcano, to form a cryptdome, "Roof Mountain". After a long period of violent explosions, the extrusion of highly viscous dacite lava continued until September 1945. As a result, this gave birth to a new mountain, "Showa-Shinzan", with a height of 407 m. During the volcanic activity, seismic observations, leveling surveys and geological-petrological investigations were carried out by Minakami, Ishikawa and Yagi (1951) and other researchers.

Figure 9.10 shows the above-mentioned seismogram recorded by the low sensitive seismometers, at the foot of the new mountain (Minakami et al., 1951). In this seismogram, the following features are noticed: (1) The amplitudes and the waveforms of these earthquakes are nearly similar. (2) Usually a number of smaller earthquakes are recorded together, but no smaller earthquakes are recorded in this case. The magnitude–frequency relation in such earthquake groups is not expressed by the Ishimoto-Iida equation (Chapter 6). Such characteristics of these earthquakes can be explained by a slick-slip mechanism.

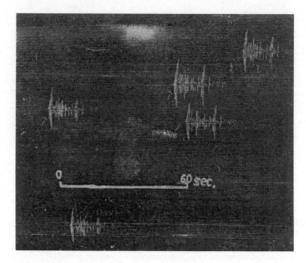

Figure 9.10. The earliest seismograms of typical characteristic earthquakes, caused by stick-slips during the new dacite lava dome formation in the 1944–1945 eruption of Usu Volcano, Hokkaido, Japan (Minakami, Ishikawa and Yagi, 1951).

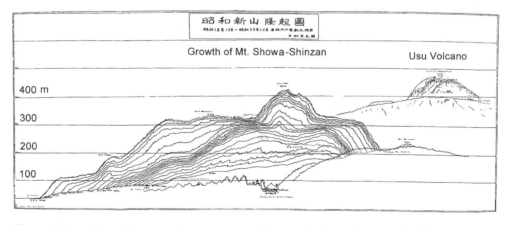

Figure 9.11. Morphological development of the newly formed mountain Showa-Shinzan recorded by Mimatsu from May 1944 to September 1945. This figure is called the "MIMATSU DIAGRAM". (Mimatsu, 1962).

During the extrusion of highly viscous magma, frictional sliding along existing cracks within the dome probably occurred, and so stick-slip events occurred successively. Masao, Mimatsu the postmaster of the Post Office near the new mountain, recorded visually the process of the uplift of the new mountain by careful sketches. Figure 9.11 shows his figure which is called the "Mimatsu Diagram". Figure 9.12 shows Usu Volcano before and after the growth of the new mountain (after Mimatsu, 1962).

In the 1977–1978 eruptions of Usu Volcano, a number of similar characteristic earthquakes also occurred during the magma extrusion process (Okada et al., 1981),

Figure 9.12. Usu Volcano before and after the growth of the new mountain (after M. Mimatsu).

and Okada (1983) discussed that these earthquakes might occur by the stick-slip mechanism.

As shown in Fig. 9.1, Unzen Volcano is located on Kyushu island, south-western Japan. During the 1990–1995 eruption of Unzen Volcano, the seismic activity increased in association with the growth of a dacite lava dome, and a number of the characteristic earthquake sequences, in which the earthquakes in each group have a similar waveform and a nearly constant magnitude, successively occurred (Shimizu et al., 2002; JMA, 1996). As a typical example, the seismogram on 12 March, 1993 obtained by JMA is reproduced in Fig. 9.13. This is essentially similar to that in Fig. 9.10. In this figure, it is interesting in particular that seismic events frequently occurred with a nearly constant time interval, although this regularity did not continue for a long time. These features strongly support the idea that these small earthquakes occurred by the stick-slip mechanism.

It should be noted that the above-mentioned characteristic earthquakes occurred during the extrusion process of highly viscous magma, such as in Usu Volcano and Unzen Volcano mentioned above. In other many volcanoes, such as andesitic volcanoes (for example, Asama Volcano and Sakurajima Volcano), many earthquakes related to volcanic eruptions occur frequently and the magnitude–frequency relations of these volcanic earthquakes can be generally expressed by the Ishimoto–Iida relation (also the Gutenberg-Richter formula). These volcanic earthquakes are mainly caused by fracture.

9.3   SANRIKU-OKI AND TOKAI-NANKAI REGIONS, JAPAN

The location of the "Sanriku-oki" region mentioned here is shown in Fig. 9.1. Matsuzawa et al. (2002) found the characteristic small-earthquake sequence off Sanriku, northeastern Honshu, Japan. According to their careful research about the occurrence of earthquakes in this region, eight similar earthquakes continued to occur

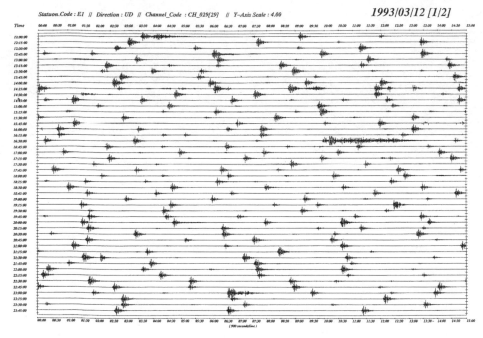

Figure 9.13.   Seismograms of small earthquakes with similar waveforms and amplitudes which occurred by stick-slip association with growth of a dacite lava dome during the 1990–1995 eruption of Unzen Volcano, southwestern Japan (after Japan Meteorological Agency, 1996).

very regularly: the magnitude is almost the same (M 4.8), the waveform is quite similar, and their epicenters locate within a very limited region. Figure 9.14 shows the M–T graph from 1956 to 1999. The mean recurrence interval of these main earthquakes (M $\approx$ 4.8) was 5.35 years and its standard deviation was 0.53 yr. These events show the same low-angle thrust fault focal mechanism and their source sizes are estimated to be around 1 km. They pointed out that the slip of each event is comparable to the cumulative amount of relative plate motion in the inter-seismic period. Therefore, it may be reasonably accepted that the occurrence of this very regular earthquake sequence can be explained by a stick-slip mechanism along the upper boundary plane of the Pacific Plate subducting at the Japan deep sea trench. In California, such characteristic small earthquake repeaters were also found in Stone Canyon (Ellsworth, 1995) and Parkfield (Nadeau and McEvilly, 1997).

In the above-mentioned examples, small characteristic earthquake sequences are discussed. Below, an example of a characteristic earthquake sequence on a large scale is explained. In Fig. 9.15, locations of rupture zones along the Nankai-Suruga trough in western Japan, which is the northern boundary of the Philippine Sea Plate, during the last three hundred years are shown (Mogi, 1985). In 1707, the whole region ruptured simultaneously. In 1854 and 1944, first the eastern part ruptured, and then

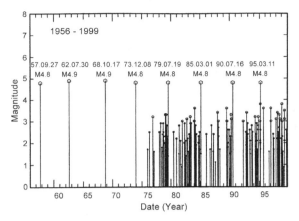

Figure 9.14. Similar earthquakes of M 4.8 which occurred regularly from 1957 to 1995 in the small region on the plate boundary off Sanriku, northeastern Japan (Matsuzawa, Igarashi and Hasegawa, 2002). Such regular occurrence of earthquakes is attributed to stick-slip of a small asperity.

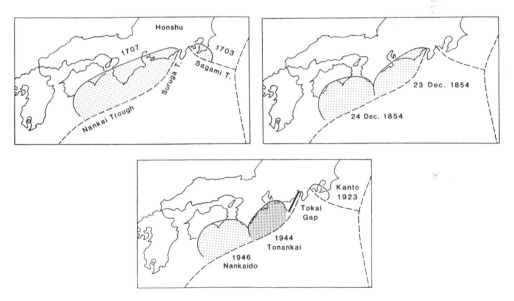

Figure 9.15. Focal regions of the three large earthquake groups (1707, 1854, 1944 and 1946) that occurred along the Nankai Trough successively, showing that Tokai region forms a seismic gap at present (Mogi, 1985).

the western part ruptured. Now, the Tokai region along the Suruga trough remains as a seismic gap. Figure 9.16 shows the M–T graph of large earthquakes along the Nankai trough in western Japan over six hundreds years (Mogi, 1974). This figure indicates that great earthquakes of M 8 class occurred repeatedly and with a high regularity. The mechanism of the earthquake sequence may be attributed to stick-slip of the large-scale asperities on the upper boundary of the subducting Philippine Sea Plate.

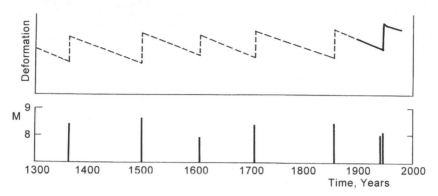

Figure 9.16.   Great earthquakes of M 8 class occurred repeatedly in the Nankai-Tokai region along the Nankai Trough with a high regularity (Mogi, 1974).

9.4   STICK-SLIP AND FRACTURE AS AN EARTHQUAKE MECHANISM

As mentioned in Chapter 4, I pointed out the high similarity between the fracture phenomena of *heterogeneous* brittle materials and natural earthquake phenomena and so I proposed the idea that earthquakes occur by fractures in the earth which is heterogeneous in structure (e.g. Mogi, 1962, 1967). On the other hand, Brace and Byerlee (1966) proposed stick-slip in frictional sliding as the mechanism of earthquakes.

In the preceding sections, typical examples of characteristic earthquake sequences which occurred by stick-slip mechanisms are shown. One of the most distinctive features of this kind of earthquake is that their magnitude–frequency relation is not expressed by the Gutenberg-Richter formula (or Ishimoto-Iida relation) which is essentially a very important characteristic of general earthquakes. In the above-mentioned examples, earthquakes with nearly a constant magnitude continued to occur. The events in stick-slip experiments show a similar feature.

In Fig. 9.17, the magnitude–frequency relation of aftershocks of the 1959 Teshikaga earthquake of M 6.3, which occurred in eastern Hokkaido, is shown after Matsumoto (1959). This graph shows that the relation is quite linear and so it is well expressed by the Ishimoto–Iida equation. In and near the focal region of the Teshikaga earthquake, geologists did not find any significant fault structures at that time. Generally it is unquestionable that most of aftershocks occurred by new fractures caused by stress redistribution due to the occurrence of the main shock. This is a typical case in which the occurrence of earthquakes cannot be explained by a simple stick-slip mechanism along pre-existing faults.

Figure 9.18 shows schematically the relation between the mechanical structures in the earth's crust and different failure mechanisms (Mogi, 1978). In cases (a) and (b), the active faults slide by slow slip or stick-slip mechanisms. In cases (c) and (d), pre-existing faults are not clear and the mechanical structure of the earth's crust is complex. In such cases, earthquakes occur mainly by fracture. Roughly, the active plate boundaries may be similar to the type on the left and the inter-plate region may be

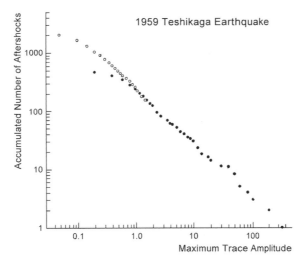

Figure 9.17. Magnitude–frequency relation of aftershocks (Ishimoto-Iida expression) of the 1959 Teshikaga earthquake (M 6.3). *m* value: 1.91 (Matsumoto, 1959).

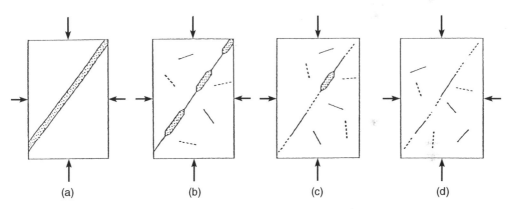

Figure 9.18. Schematic diagram illustrating different mechanical structures in the earth's crust in which sliding, stick-slips and fractures occur.

similar to the type on the right. When the time interval between earthquakes is long, the fault surface may heal with time, and so earthquakes may occur by fracture in this case.

In conclusion, there are various mechanisms of earthquake occurrence. Types of failure of the earth's crust, such as slow slip, stick-slip and fracture, may depend on various factors, such as mechanical structures, and deformation rates and various physical properties of pre-existing fault planes.

REFERENCES

Amelung, F., S. Jónsson, H. Zebker and P. Segall. (2000). Widespread uplifting on Galápagos Volcanoes observed with radar interferometry. Nature, Vol. **407**, 993–996.

Brace, W. F. and J. D. Byerlee. (1966). Stick-slip as a mechanism for earthquakes. Science, **158**, 990–992.

Ellsworth, W. L. (1995). Characteristic earthquakes and long-term earthquake forecasts: implications of central California seismicity, in *Urban Disaster Mitigation: The role of Science and Technology*, edited by F. Y. Chen and M. S. Sheu, Elsevier, Oxford.

JMA. (1996). Unzen-Fugendake Volcanic Activity, Memoirs of Fukuoka Meteorological Observatory, No. 51, 315pp.

Matsumoto, T. (1959). On the spectral structure of earthquake waves. Bull. Earthquake Res. Inst., Tokyo Univ., **37**, 265–277.

Matsuzawa, T., T. Igarashi and A. Hasegawa. (2002). Characteristic small-earthquake sequence off Sanriku, north eastern Honshu, Japan. Geophys. Res. Lett., Vol. 29, No. 11, 10.1029/2001 GLO14632.

Mimatsu, M. (1962). Showa-Sinzan Diary: Complete records of observation of the process of the birth of Showa-Sinzan, Usu Volcano, Hokkaido, Japan, 179pp.

Minakami, T., T. Iishikawa and K. Yagi. (1951). The 1944 Eruption of Volcano Usu in Hokkaido, Japan. Bull. Volcanol., **11**, 45–157.

Miyabe, N. (1934). Deformation of the earth's crust in the neighbourhood of Sakurazima. Bull. Earthquake Res. Inst., Tokyo Univ., **12**, 471–481. (in Japanese).

Nadeau, R. M. and T. V. McEvilly. (1999). Fault slip rates at depth from recurrence intervals of repeating microearthquakes. Science, 285, 718–721.

Mogi, K. (1957). On the relation between the eruptions of Sakurazima Volcano and the crustal movements in its neighbourhood. Bull. Volcanol. Soc. Japan, **1**, 9–18. (in Japanese).

Mogi, K. (1958). Relations between the eruptions of various volcanoes and the deformations of the ground surfaces around them. Bull. Earthquake Res. Inst., Tokyo Univ., **36**, 99–134.

Mogi, K. (1962). Study of elastic shocks caused by the fracture of heterogeneous materials and its relation to earthquake phenomena. Bull. Earthquake Res., Inst., Tokyo Univ., **40**, 125–173.

Mogi, K. (1967). Earthquakes and fractures. Tectonophysics, **5**, 35–55.

Mogi, K. (1974). Earthquakes as fractures in the earth. Proc. 3rd Congr. Int. Soc. Rock Mech., **1**, Part A, 559–568.

Mogi, K. (1978). Rock mechanics and earthquakes. In: Physics of Earthquakes (ed. H. Kanamori), Iwanami-Shoten, Tokyo, 211–262. (in Japanese).

Okada, H. (1983). Comparative study of earthquake swarms associated with major volcanic activities. In Arc Volcanism: Physics and Tectonics, (Shimozuru, D. and Yokoyama, I., eds.), 43–61. Terra Scientific Publishing Company, Tokyo.

Okada, H., H. Watanabe, H. Yamashita, and I. Yokoyoma. (1981). Seismological significance of the 1977–1978 eruptions and the magma intrusion process of Usu Volcano, Hokkaido. J. Volcanol. Geotherm Res., **9**, 311–334.

Ōmori, F. (1914–1922). The Sakurajima eruptions and earthquakes. Bull. Imperial Earthquake Investigation Committee, **8**, 525pp.

Ryan, M. P., J. Y. K. Blevins, A.T. Okamura and R. Y. Koyanagi. (1983). Magma reservoir subsidence mechanics: Theoretical summary and application to Kilauea Volcano, Hawaii. J. Geophys. Res., Vol. 88, No. B5, 4147–4181.

Shimizu, H., K. Umakoshi, N. Matsuwo and K. Ohta. (1992). Seismological observations of Unzen Volcano before and during the 1990–1992 eruption. In: Unzen Volcano, the 1990–1992 eruption (Yanagi, T., Okada, H. and Ohta, K. eds.) 38–43. The Nishinippon & Kyushu Univ. Press, Fukuoka.

Tsuboi, C. (1929). Block movements as revealed by means of precise levellings in some earthquake districts of Japan. Bull. Earthquake Res., Tokyo Univ., **7**, 103–114.

Yamakawa, N. (1955). On the strain produced in a semi-infinite elastic solid by an interior source of stress. J. Seismol. Soc. Japan, **8**, 84–98. (in Japanese).

CHAPTER 10

Some features in the occurrence of recent large earthquakes

10.1 GLOBAL PATTERN OF SEISMIC ACTIVITY

Earthquakes do not occur around the earth at random, but are mainly concentrated in limited belt-like zones. Figure 10.1 shows the epicentral distribution of large shallow earthquakes ($M \geq 7.5$, focal depth ≤ 40 km) in about 100 years from 1903 to 2005. The five numbered large circles are the great earthquakes of M_w greater than 9.0. The first instrumental magnitude scale M was proposed by Richter (1935). However, M was proposed to apply to moderate-scale earthquakes and focus on relatively short-period seismic waves, so it is not appropriate as a quantitative scale for great earthquakes of M 8–9 class. Kanamori (1977) proposed the new magnitude scale M_W, using seismic moment based on the fault length and amount of displacement. Magnitude M_W and M are nearly identical for moderate earthquakes with a fault length 100 km or less, so the conventional Richter magnitude expresses an appropriate magnitude for such earthquakes. In this book, M is used to express the Richter magnitude for earthquakes of M 7.5–8 or smaller and the moment magnitude M_W for earthquakes of greater magnitude.

In Fig. 10.1, the seismicaly active areas around the earth can be classified as follows.

(a) Circum – Pacific seismic zone

The rim of the Pacific Ocean is one of the most active seismic zones in the world. These earthquakes occur by the continuous movement of oceanic plates toward the continental plate along the plate boundaries, where plates collide with and subduct under the continents. The west coast of the North American continent is the site of a transform fault where the two sides of the plate boundary move sideways with respect to each other.

(b) Alps – Himalaya – Sunda seismic zone

Another active seismic zone is the belt running from the Sunda Islands (including Sumatra and Java) through Burma, the Himalaya, Iran and Turkey to Greece and Italy. Sumatra and Java have virtually the same subduction characteristics as the main parts of the circum – Pacific seismic zone, but further westward the degree of activity of this seismic zone decreases slightly and is somewhat spread over a wider area. This seismic activity is caused by the movement of the Australian (and Indian) Plate in the NNE direction relative to the Eurasian Plate.

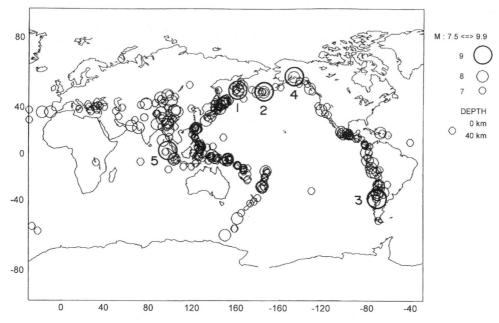

Figure 10.1. Epicentral distribution of large shallow earthquakes (M ≥ 7.5, focal depth ≤40 km) in about 100 years from 1903 to 2005 in the world. Largest circles : M_W ≥ 9.0. (Modified Ishikawa, 2004; Mogi, 2005).

(c) Continental seismic regions

There is considerable seismic activity on the Chinese mainland; these earthquakes occur within continents and they are referred to as intra-plate earthquakes or continental earthquakes. It is noteworthy that the seismic region is surrounded by active plate boundaries. It was pointed out that there was a close relation between the activity in the continent and the seismicity in the surrounding plate boundaries (e.g. Mogi, 1993).

The next problem is the temporal variation of the seismic activity around the whole earth. Figure 10.2 (a) shows the M–T graph of large shallow earthquakes (M ≥ 8, focal depth ≤40 km). The vertical axis indicates M_W and solid circles indicate great earthquakes of M_W larger than 9.0. Since the seismic energy is released mainly by these large earthquakes, we can see the temporal change of seismic activity around the whole earth from this graph. This result shows that the 15 years from 1950 to 1965 was the most active period, and then the activity decreased and again began to increase from 1995. During this activation process, the great 2004 Sumatra earthquake occurred.

Figure 10.2 (b) shows the temporal variation of the number of people killed by earthquakes [Data from Utsu (1999) and recent data from USGS]. Kanamori (1978) reported the graphs for the period (1900–1980) similar to those of Fig. 10.2 (a) and (b). Generally, it is expected that the temporal changes in seismic activity and earthquake damage (the number of people killed by earthquakes) vary in parallel. However, their

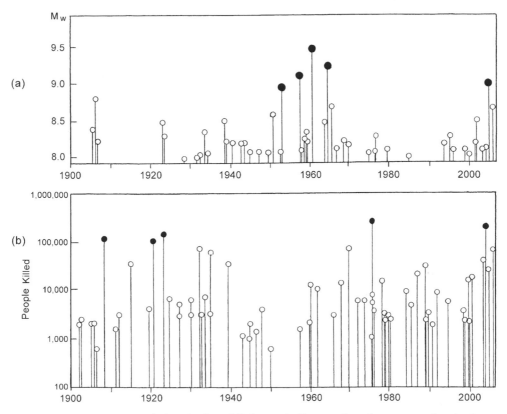

Figure 10.2.   (a) Magnitude (M_W) of world's large shallow earthquakes greater than is shown against time in the past about 100 years. (b) Number of people killed by earthquakes in the world is shown against time. [Data from Utsu (2004), report from USGS (2005), and information from related governments.]

temporal changes are not similar, but the trends are virtually reversed in the period (1950–1965). How are these contradictory results to be explained?

To understand this result, I examined the epicentral distribution of large shallow earthquakes shown in Fig. 10.1 (Mogi, 1979). Figure 10.3 shows the temporal changes in seismicity in high-latitude regions (40° and above) and low-latitude regions (below 40°). In order to express the degree of seismic activity, both magnitudes M_W and M of large shallow earthquakes are indicated, but they show the same trend. The population density of the high-latitude regions is low, so despite the fact that great earthquakes of M_W 9 class occurred during the period (1950–1965), fortunately there have been few deaths. The population is very dense in the lower-latitude regions, however, so the loss of life has been great even in the case of smaller earthquakes. Thus, it is understandable that human losses shown in Fig. 10.3 correspond more or less to changes in the degree of seismic activity in the lower-latitude regions. From these results it can be said that the lower-latitude regions in recent times are currently in an active period, with earthquake disasters occurring one after another. Since 1970, the level of earthquake disaster is high, particularly with the 1976 Tangshan (China) earthquake (M 7.6, 39.6°N, 240,000 deaths), the 2004 Sumatra (Indonesia) earthquake (M_W 9.1–9.3, 3.3°N, about 220,000 deaths), and the 2005 Pakistan earthquake (M 7.7, 34.5°N, about 87,000 deaths).

Figure 10.4 shows the space–time distribution of large shallow earthquakes greater than M 7.8 in the world (1901–2005.3). The ordinate is latitude and the abscissa is

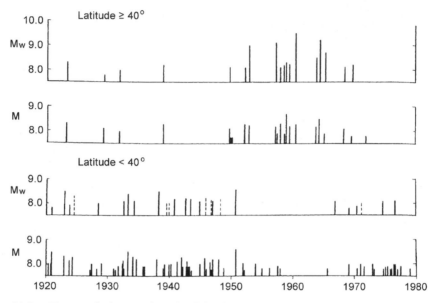

Figure 10.3.   Temporal changes in seismicity in high-latitude regions (40° and above) and low-latitude regions (below 40°). In order to express the degree of seismic activity, both magnitudes M_W and M of large, shallow earthquakes are indicated, but they show the same trend. (Mogi, 1979).

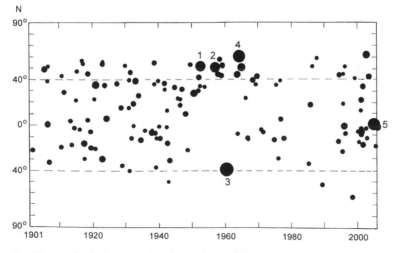

Figure 10.4.   Space–time distribution of large shallow earthquakes ($M_W \geq 7.8$) in the world (1901–2005). (Modified Mogi, 1979; Utsu, 2004). Largest circles (1–5), $M_W \geq 9.0$.

time. Large solid circles with numerals indicate great shallow earthquakes greater than M_W 9.0. In this figure, it can be seen clearly that during the above-mentioned period (1950–1965), the low-latitude region (40°N–40°S) was quiet and the high-latitude regions (north and south) were very active. Although the mechanism of this change of seismic activity is not clear, this change suggests that some global events occurred within the earth. This problem should be investigated on a global basis. And it is also noted that the great 2004 Sumatra earthquake occurred during the active period of the whole earth.

10.2   ACTIVE AND QUIET PERIODS IN THE MAIN SEISMIC ZONES

10.2.a   *Alaska – Aleutian – Kamchatka – N. Japan*

There is sometimes some kind of regularity in temporal variations in the degree of activity in the main world seismic zones (Mogi, 1974). Figure 10.5 shows the distributions of focal regions of great shallow earthquakes ($M_W \geq 8.0$) along the plate boundary through Alaska, the Aleutians, Kamchatka, the Kuriles and northern

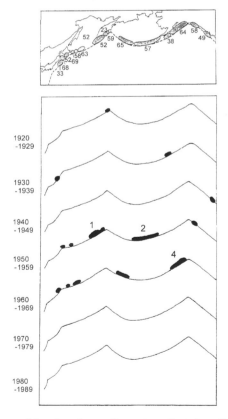

Figure 10.5.   Distributions of focal regions of great earthquakes ($M_W \geq 8.0$) along the plate boundary from Alaska to N. Japan at 10-year intervals. (Modified Mogi, 1993).

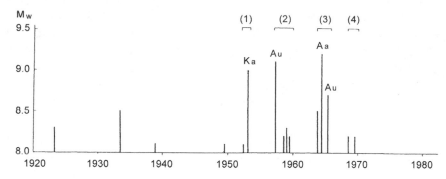

Figure 10.6.   M–T graph of great shallow earthquakes ($M_W \geq 8.0$) in the Alaska – Aleutian – Kamchatka – Kurile – northern Honshu seismic belt. (Mogi, 1993).

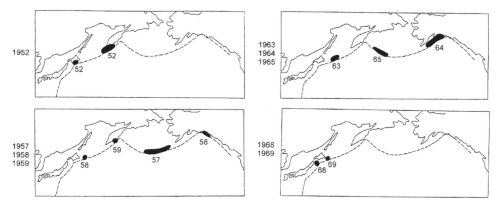

Figure 10.7.   Locations of great earthquakes of 4 groups that occurred near each other in time in the northern circum-Pacific seismic belt. These 4 groups are labeled (1) to (4) in Fig. 10.6. (Mogi, 1993).

Japan at 10-year intervals starting in 1920 (modified from Mogi, 1993). In this figure, it can be seen that the period (1950–1970) was a very active period, in which three great earthquakes of M_W 9.0 and above occurred, and the other periods were quiet. In the active period, the ruptures occurred along nearly the whole length (6000 km) of this plate boundary between the Pacific and North American Plates.

Figure 10.6 shows the M–T graph of great earthquakes of M_W 8.0 and larger in this seismic zone. Earthquakes in the period (1950–1970) can be divided into 4 groups that occurred near each other in time. These are labeled (1) to (4) in Fig. 10.6. The locations where the earthquakes in each of these 4 groups occurred are shown in Fig. 10.7. The numerals indicate the time of the occurrence of these earthquakes. It seems that each great earthquake led to the occurrence of another earthquake along the same plate boundary.

In addition, this long plate boundary is almost completely covered by the rupture zones of these great earthquakes that occurred in a relatively short time ($\sim$20 years).

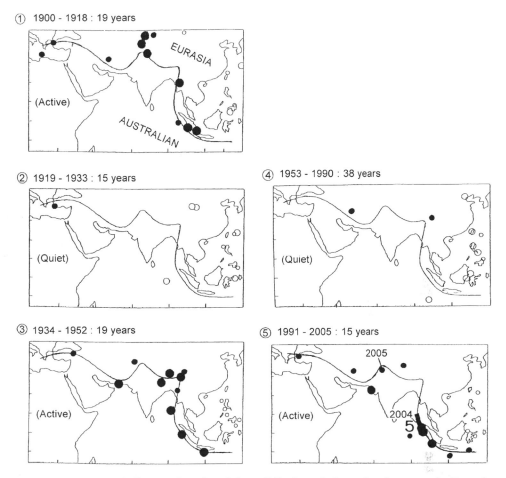

Figure 10.8. Large shallow earthquakes ($M_W \geq 7.7$) along the boundary between the Eurasian Plate and the Australian Plate (the Alps – Himalaya seismic belt) are plotted at various time intervals. Data are from U.S. Coast and Geodetic Survey. (Modified from Mogi, 1974).

This result suggests that some stress change, such as a few MPa (see e.g. Kanamori and Anderson, 1975), may occur throughout the whole plate boundary in this region. It is an interesting problem that such successive occurrence of great earthquakes along the very long plate boundary may cause a rhythmical change in the plate motion (Mogi, 1993).

10.2.b *Alps – Himalaya – Sunda*

Thirty years ago, large shallow earthquakes ($M \geq 7.7$) along the boundary between the Eurasian Plate and the Australian Plate (African Plate along one part) were plotted at intervals of about 20 years. It was found that the active and quiet periods repeat (Mogi, 1974). Figure 10.8 is a similar figure for 1900 to 2005. The repeatition of the

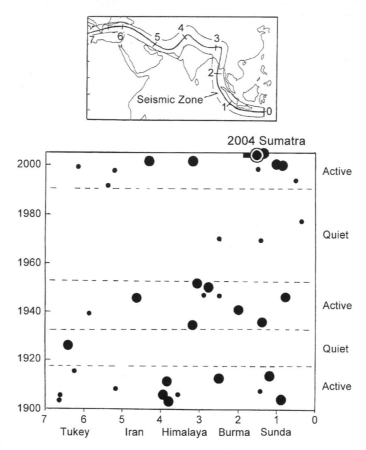

Figure 10.9.   Space–time distribution of large shallow earthquakes in the Alps – Himalaya – Sunda seismic zone in the past 100 years. The double solid circle indicates the 2004 Sumatra earthquake. Large solid circles indicate large shallow earthquakes of $M \geq 8.0$, small solid circles $8.0 > M \geq 7.7$. (Data from Utsu (1999) and PDE.)

active and quiet periods can be also seen. However, the quiet period from 1953 is long and continued for 38 years. In the following period from 1991, the activity increased and the 2004 great Sumatra earthquake of M_W 9.0 (~9.3) occurred. This shows how this huge seismic zone, more than 15,000 km long, becomes active all at once, and then suddenly becomes inactive. Figure 10.9 shows the space – time distribution of large – shallow earthquakes ($M \geq 7.7$) in the Alps-Himalaya-Sunda seismic zone (the boundary between the Eurasia and the Indo-Australian plate). The horizontal axis is the length along the plate boundary (not linear). It is remarked that the seismic active period and the quiet period repeated simultaneously in this very long plate boundary. Fig. 10.10 shows the M–T graph of these large shallow earthquakes ($M \geq 7.7$) in the same seismic zone. In these figures, it is pointed out that the 2004 great Sumatra earthquake was preceded by a long quiet period and then an activated period of 12 years. In 2005, the Pakistan earthquake (M 7.7) occurred and a great number of people were killed.

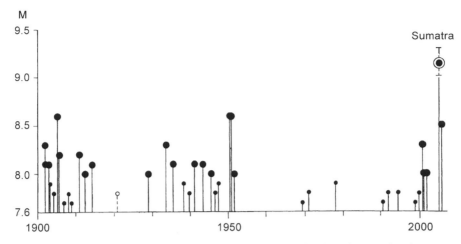

Figure 10.10. Temporal variation of the seismic activity in the Alps – Himalaya – Sunda seismic zone shown by a M–T graph. The 2004 Sumatra earthquake was preceded by a markedly long quiet period and then an activated period. The magnitude of the Sumatra earthquake is determined to be somewhat different by different researchers, as shown in this figure. (Data from Utsu (1999) and PDE.)

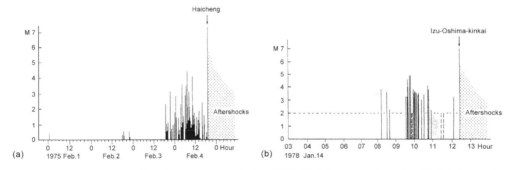

Figure 10.11. (a) Foreshock sequence, of the 1975 Haicheng earthquake (Raleigh et al., 1977). (b) The 1978 Izu-Oshima-kinkai earthquake. (Data from JMA catalogue; Mogi, 1982). (Mogi, 1987).

10.3 SOME PRECURSORY SEISMIC ACTIVITY OF RECENT LARGE SHALLOW EARTHQUAKES

10.3.a *Introduction*

Seismicity changes prior to a large earthquake would be an important clue in predicting the occurrence of a large earthquake. Actually, there are numerous cases in which there is an increase in the number of small earthquakes immediately before a large earthquake. In Chapter 4, the foreshocks of the 1930 Kita-Izu earthquake (M 7.3) are presented. In Fig. 10.11, two other typical examples of major earthquakes preceded by marked foreshock activities are shown. (a) is the case of the 1975 Haichen

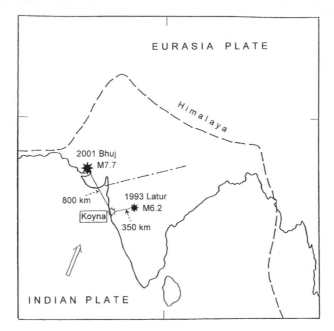

Figure 10.12.   Location of the 2001 Bhuj (India) earthquake (M 7.7) and the 1993 Latur earthquake (M 6.2), and the Koyna earthquakes which have continued to occur in the Koyna dam region since 1963. The Koyna dam is 350 km from Latur and 800 km from Bhuj. The chain line indicates Narmada Son tectonic line. (Mogi, 2001).

(China) earthquake of M 7.3 compiled by Raleigh et al. (1977). This earthquake was successfully predicted on the basis of various precursory phenomena in which the occurrence of remarkable foreshocks was most important. (b) is the case of the 1978 Izu-Oshima-Kinkai (Japan) earthquake of M 7.0. Data are from JMA catalogue. In this case also, various precursory phenomena were observed. There are numerous cases in which there is an increase in the number of small earthquakes immediately before a large earthquake in its epicentral region. In this section, the cases in which the seismic activity increased prior to great earthquakes in distant regions are explained.

10.3.b   *2001 Bhuj (India) earthquake*

On 26 January, 2001, the Bhuj earthquake of M_W 7.7 occurred in western India, as shown in Fig. 10.12. About 25,000 people were killed by this large earthquake.

On September 30, the 1993 Latur earthquake of M 6.2 occurred at the center of the Indian Peninsula (see Fig. 10.12). In the next year, a foreign Advisory Team (A.C. Johnston, Max Wyss and Mogi) visited India for the period Jan. 30 to Feb. 17, 1994 under the sponsorship of UNDHA (United Nations Department of Humanitarian Affairs) and UNDP (United Nations Development Program). In this team, I researched particularly the space-time distribution of earthquakes in India on the basis of the Earthquake Catalogue of the India Meteorological Department, modified by

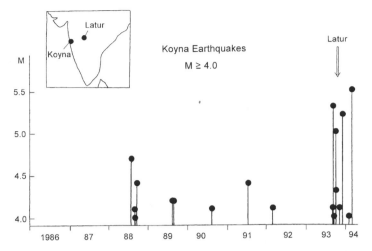

Figure 10.13.   M–T graph of the Koyna (dam) earthquakes (M ≥ 4.0) and the Latur earthquake (6.2). The Koyna earthquakes occurred just before and after the Latur earthquake. Earthquake data from Catalogue of India Meteorological Department modified by National Geophysical Research Institute, India. (Mogi, 1994).

the National Geophysical Research Institute, India. At Koyna, situated near the west coast of the Indian Peninsula, a large dam was constructed in 1962 and dam-induced earthquakes began to occur in 1967. In December 1967, the largest earthquake of M 6.3 occurred and then smaller earthquakes continued to occur in the Koyna region. During my stay in India, I examined the temporal variation of the seismic activity in the Koyna region, before the 1993 Latur earthquake of M 6.2, although the Koyna dam is distant ($\Delta = 350$ km) from the epicenter of the Latur earthquake. I supposed that the Koyna region might be a special region which was very sensitive to a change of tectonic stress because the fracture strength of the crust in this region was extremely weakened by the induced water, and so the occurrence of a distant major earthquake might have influence on the seismic activity in the Koyna region. Particularly, the Indian Peninsula is a part of the Indian Plate which is mechanically uniform, so such influence of events in distant regions may be possible.

Figure 10.13 shows the M–T graph of Koyna (dam) earthquakes of M ≥ 4.0 and above. Data are taken from the Earthquake Catalogue of the India Meteorological Department modified by the National Geophysical Research Institute (Director, H. K. Gupta). In this figure, the 1993 Latur earthquake is indicated by an arrow. This figure shows a close relation between the Latur earthquake and the Koyna earthquakes. Particularly, it is noted that the activity in the Koyna region increased immediately prior to the Latur earthquake (Mogi, 1994).

The 2001 Bhuj earthquake of M 7.7 occurred in western India. The distance (Δ) between the Bhuj earthquake and the Koyna regions is about 800 km, as shown in Fig. 10.12. Usually it may be thought that there is no relation between them. However, in consideration of the case of the Latur earthquake, I examined the relation between the Bhuj earthquake and the seismic activity in the Koyna region.

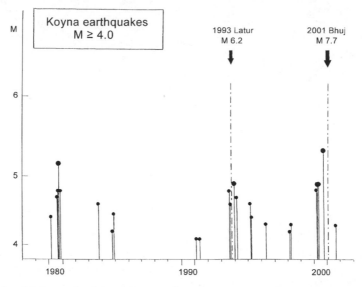

Figure 10.14.   M–T graph of the Koyna earthquakes (M ≥ 4.0) from 1978 to 2001. The 1993 Latur earthquake (M 6.2) and the 2001 Bhuj earthquake (M 7.7) are indicated by thick arrows. The relation between the Koyna earthquakes and the two major earthquakes is noticeable. Earthquake data from ISC and USGS (PDE) catalogues. (Mogi, 2001).

Figure 10.14 shows the M–T graph of the Koyna earthquakes of M ≥ 4.0 and above. Data are taken from ISC (International Seismological Center) and USGS (PDE). (Mogi, 2001). The thick arrows indicate the 1993 Latur earthquake (M 6.2, $\Delta = 350$ km) and the 2001 Bhuj earthquake (M 7.7, $\Delta = 800$ km). In this figure, a noticeable relation between the Koyna earthquakes and the 2001 Bhuj earthquake can be seen. In particular, a significant precursory seismic activity can be recognized also in the case of the Bhuj earthquake.

The above-mentioned results suggest that the Koyna dam-induced earthquakes may be an indicator of the tectonic stress state in a wide surrounding region.

10.3.c   *2003 Tokachi-oki (Japan) earthquake*

On September 26, the 2003 Tokachi-oki earthquake of M 8.0 occurred off the south-east coast of Hokkaido, Japan. This earthquake was a low-angle, thrust-type, inter-plate earthquake which occurred on the upper surface of the Pacific Plate subducting beneath Hokkaido. The location, the focal mechanism and the magnitude are quite similar to those of the 1952 Tokachi-oki earthquake of M 8.2. In 1973, the author pointed out that the 1952 Tokachi-oki earthquake and several other large shallow earthquakes were preceded by marked deep seismic activity (Mogi, 1973). Since the 2003 Tokachi-oki earthquake reoccurred at a 50-year interval, now we can examine the author's supposition. In this section, the relation between great shallow earthquakes and deep focus earthquakes in and around Japan is discussed (Mogi, 2004, 2005).

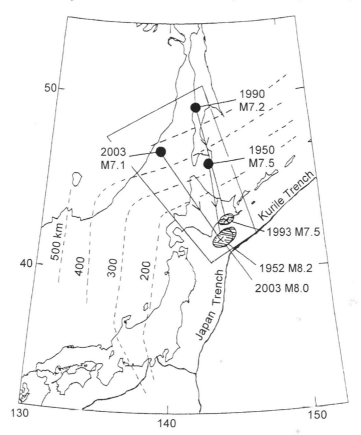

Figure 10.15.   Locations of the focal regions of the 2003 Tokachi-oki earthquake, the 1952 Tokachi-oki earthquake and the 1993 Kushiro-oki earthquake, and the largest deep focus earthquakes prior to these large shallow earthquakes which occurred in the rectangular region shown in this figure. Solid curve: deep sea trench; broken line: equal-depth line of the deep seismic plane. (Mogi, 2004).

Figure 10.15 shows the area in and around the Japanese islands with the Kurile Trench and the Japan Trench, and the focal regions of the 2003 Tokachi-oki earthquake, the 1952 Tokachi-oki earthquake and the 1993 Kushiro-oki earthquake. The solid circles are the largest deep-focus earthquakes prior to these large shallow earthquakes. In this section, *shallow* is used for focal depth ≤ 100 km and *deep* is for focal depth 200–600 km. The broken curves indicate the equal-depth contours of the deep seismic plane.

Figure 10.16 (a) shows epicentral locations of earthquakes ($M \geq 5.0$) which occurred from 1978 to 2003 in and around Hokkaido and Fig. 10.16 (b) shows the vertical projection of earthquakes in the rectangular region in (a) along line AB. The deep seismic plane can be seen clearly.

Figure 10.17 shows M–T graphs of the deep-focus earthquakes (solid circles: $M \geq 5.0$, focal depth ≥ 200 km) which occurred in the rectangular region indicated

1978 1/1 0:0 - 2003 12/31 24:0

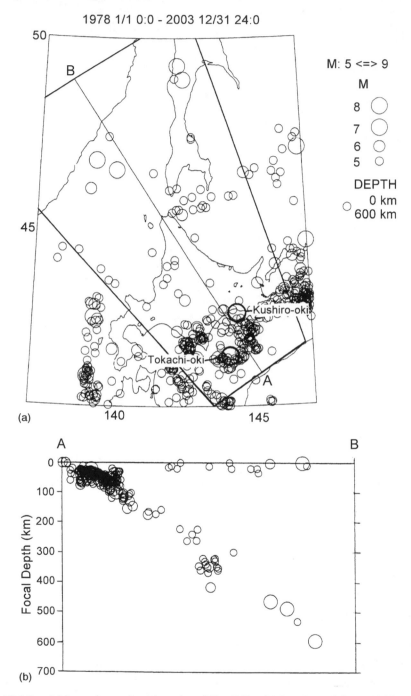

Figure 10.16.   (a) Locations of earthquakes (M ≥ 5.0) which occurred from 1978 to 2003 in and around Hokkaido. Large circles indicate the 2003 Tokachi-oki earthquake (M 8.0) and the 1993 Kushiro-oki earthquake (M 7.5). Data are from JMA catalogue. (b) Vertical projection of earthquakes in the rectangular region in (a) along AB line. The deep seismic plane can be seen clearly.

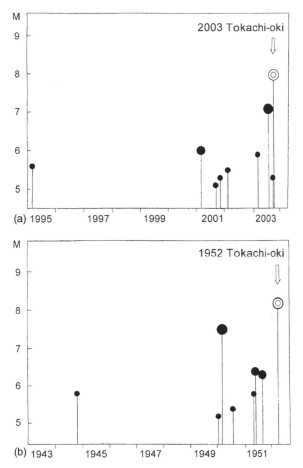

Figure 10.17. (a) M–T graph of the deep-focus earthquakes (M ≥ 5.0, focal depth ≥200 km), which occurred in the rectangular region shown in Fig. 10.15, preceding the 2003 Tokachi-oki earthquake. (b) The similar graph in the case of the 1952 Tokachi-oki earthquake.

in Fig. 10.15 and great shallow earthquakes (open circles). The seismic activity in the deep region increases prior to the great shallow earthquake in both cases of 1952 and 2003. In these two cases, the pattern was quite similar and a large deep-focus earthquake of M 7.0 or above occurred. This result strongly supports the opinion that deep-focus foreshocks occur in a certain region.

Figure 10.18 (a) and (b) show the epicentral locations of deep-focus earthquakes preceding the 2003 Tokachi-oki earthquake in the two successive periods (1995–2000) and (2001–2003.9) and the 1952 Tokachi-oki earthquake in the two successive periods (1943–1949) and (1950–1952.3), respectively. These figures also show that deep-focus earthquakes occurred immediately prior to these great shallow earthquakes.

Figure 10.19 shows the case of the 1993 Kushiro-oki earthquake (M 7.5). This earthquake occurred on January 15, 1993 at a depth of 100 km in the adjacent region. According to the observation by JMA (1993) and Hokkaido University (1993), this

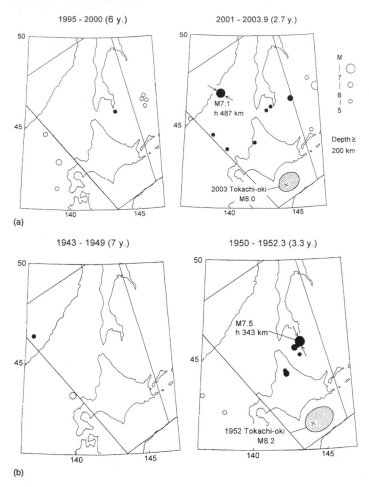

Figure 10.18.   (a) Locations of deep-focus earthquakes (M ≥ 5.0, focal depth ≥ 200 km) preceding the 2003 Tokachi-oki earthquake in the two successive periods (1995–2000) and (2001–2003.9). (b) The similar figures in the case of the 1952 Tokachi-oki earthquake in the two periods (1943–1949) and (1950–1952.3).

was an intra-plate earthquake, which occurred within the Pacific Plate subducting beneath Hokkaido from the Kurile Trench. According to the aftershock distribution and the focal mechanism analysis, the fault plane of this earthquake was nearly horizontal and the lower surface of the fault slipped northward (JMA, 1993; Kasahara, 2000) (see Fig. 10.23). Figure 10.19 (a) shows the M–T graph of deep-focus earthquakes (M ≥ 5.0, focal depth ≥ 200 km), which occurred in the rectangular region indicated in Fig. 10.15, preceding the 1993 Kushiro-oki earthquake. Figure 10.19 (b) shows the epicentral locations of the deep-focus earthquakes in the two successive periods (1978–1985) and (1986–1993.1). In this case the activity of deep-focus earthquakes also increased before the 1993 Kushiro-oki earthquake.

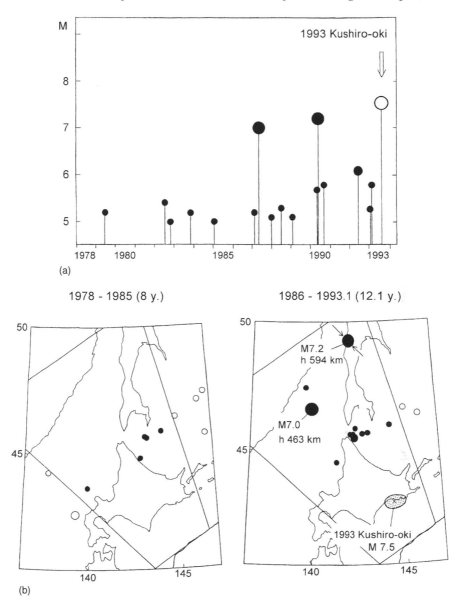

Figure 10.19.   (a) M–T graph of the deep-focus earthquakes (M ≥ 5.0, focal depth ≥200 km), which occurred in the rectangular region shown in Fig. 10.15, preceding the 1993 Kushiro-oki earthquake. (b) Locations of the deep-focus earthquakes in the two successive periods (1978–1985) and (1986–1993, 1).

When the largest deep-focus earthquake of M 7.2 occurred on May 12, 1990 beneath central Sakhalin, Motoya and Mogi discussed the possibility of the occurrence of a large shallow earthquake along the western Kurile Trench, because the 1952 Tokachi-oki earthquake was preceded by a large deep-focus earthquake, as mentioned

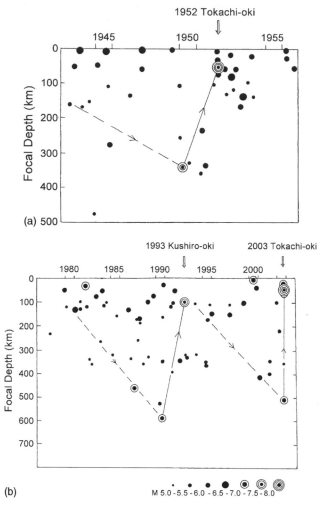

Figure 10.20.   (a) Focal depths of earthquakes before and after the 1952 Tokachi-oki earthquake in the rectangular region shown in Fig. 10.15 are plotted against time. (Modified Mogi, 1988). (b) The similar graph in the case of the 1993 Kushiro-oki earthquake – the 2003 Tokachi-oki earthquake.

above. Actually, the 1993 Kushiro-oki earthquake, which is an intra-plate earthquake, occurred. After the Kushiro-oki earthquake, deep seismic activity decreased rapidly.

Figure 10.20 (a) shows the temporal variation of the focal depths of earthquakes before and after the 1952 Tokachi-oki earthquake in the rectangular region in Fig. 10.15 and plotted against time (modified from Mogi, 1988). Following the gradual downward migration of seismic activity shown by a broken line, several deep-focus earthquakes including the largest one (M 7.5, focal depth 343 km) occurred, the rapid upward migration shown by a solid line followed and the 1952 Tokachi-oki earthquake (M 8.2, focal depth 54 km) occurred. (Small earthquakes shallower than 100 km are not plotted in Fig. 10.20).

1930 1/1 0:0 - 2004 10/30 24:0

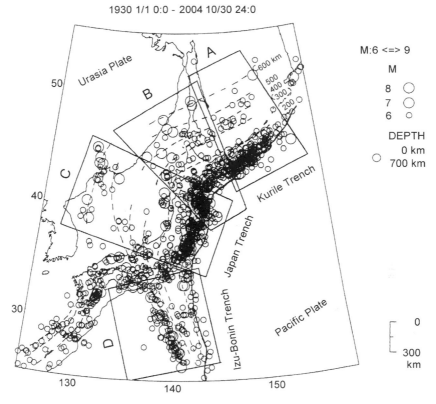

Figure 10.21.   Locations of earthquakes (M ≥ 6.0) in and around the Japanese Islands. The solid curve shows deep sea trenches and the broken curves show the equal-depth contours of deep seismic plane, which corresponds to the subducting slab. As shown in this figure, this subduction zone is divided into the four rectangular regions (A, B, C and D).

Figure 10.20 (b) shows the temporal variation of the focal depths of earthquakes which occurred during the period (1978–2003) in the same region. In this figure, two phases can be clearly distinguished. One is related to the 1993 Kushiro-oki earthquake and the other is the process related to the 2003 Tokachi-oki earthquake. It is noteworthy that the gradual downward migration of seismic activity, the occurrence of the largest deep-focus earthquake, the rapid upward migration of seismic activity and the occurrence of the large shallow earthquake can be seen in these two processes. This pattern of deep seismic activity prior to a large shallow earthquake is quite similar to that for the case of the 1952 Tokachi-oki earthquake, mentioned above. Thus, it is concluded that the above-mentioned process observed in the case of the 1952 Tokachi-oki earthquake is not accidental, but it is common to the three large shallow earthquakes which occurred successively off the south-east coast of Hokkaido.

About the above-mentioned relation between the large deep-focus earthquakes and the shallow great earthquakes, some seismologists express some doubt. They argue that there are many cases in which large deep-focus earthquakes of M 7 class

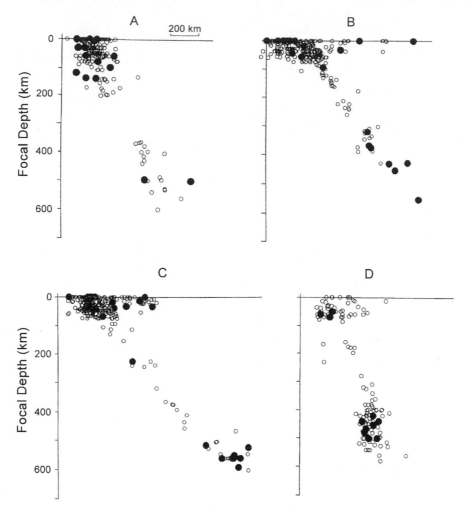

Figure 10.22.   Vertical projection of earthquakes in the deep seismic plane which shows the subducting slab. The subduction of slabs in Regions A, C and D start from the Kurile trench, Japan trench and Izu-Bonin trench, respectively. Region B where the Kurile trench meets the Japan trench may be a special region. In Region B, the seismic activity is continuously high from the shallow to deep region. Solid circles: $M \geq 7.0$; open circles: $7.0 > M \geq 6.0$. (Data from JMA catalogue, 2003).

occurred, but large shallow earthquakes did not occur. For example, large deep-focus earthquakes occur frequently along the Izu-Bonin Trench, but no great shallow earthquakes occur along this trench. I discussed this problem previously (Mogi, 1981), but I explain it more explicitly below.

Figure 10.21 shows epicentral locations of earthquakes ($M \geq 6.0$) in and around the Japanese islands. Thick solid curves indicate the Kurile Trench, Japan Trench and Izu-Bonin Trench and the broken curves indicate the equal-depth contour of the deep

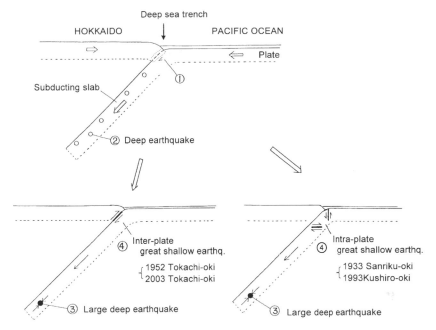

Figure 10.23.   Diagram illustrating schematically the mechanism of the occurrence of large shallow earthquakes and their relation to the deep seismic activity. Both the cases of the inter-plate earthquake (left) and the intra-plate earthquake (right) are shown. The processes ①, ②, ③ and ④ successively occur. (Mogi, 2004).

seismic plane, which corresponds to the subduction slab. Along the deep sea trench, the subduction zone is divided into four rectangular regions (A, B, C and D from north to south).

Figure 10.22 shows the vertical projections of earthquakes in the deep seismic plane which shows the subducting slab. The subduction of slabs in Region A, C and D starts from the Kurile Trench, Japan Trench and Izu-Bonin Trench, respectively. Region B where the Kurile Trench meets the Japan Trench may be a special region, and it is remarked that the seismic activity is continuously high from the shallow to deep region. In these figures, solid circles are $M \geq 7.0$ and open small circles are $7.0 > M \geq 6.0$.

This vertical projection of earthquakes suggests that the subducting slab in region B acts as a good stress guide, and the slabs in Region A, C and D do not act as mechanical stress guides. In particular, it is supposed that there is no mechanical relation between the shallow and the deep parts. On the other hand, it may be expected that the deep part has a close mechanical relation with the shallow part in region B.

Figure 10.23 shows a diagram illustrating schematically the mechanism for the occurrence of large shallow earthquakes in the subduction zone and their relation to the deep seismic activity. The cases for the inter-plate earthquakes and the intra-plate earthquakes are elucidated by this figure.

Figure 10.24 shows the relation between large deep-focus earthquakes and great shallow earthquakes in Region B during the period from 1927 to 2004 October. Solid

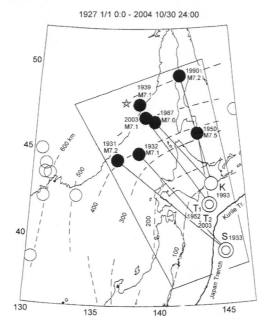

Figure 10.24.   Relation between large deep-focus earthquakes and great shallow earthquakes in Region B during the period from 1927 to October, 2004. Solid circle: deep-focus earthquakes of M ≥ 7.0 and focal depth (h) ≥ 300 km; open double circles: great shallow earthquakes of M ≥ 8.0; open circle: 1993 Kushiro-oki earthquake (M 7.5, h = 100 km).

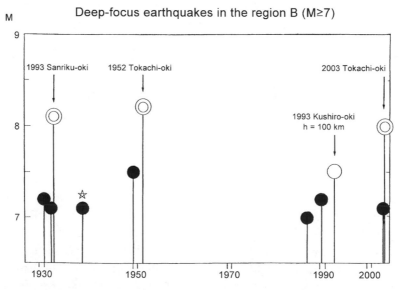

Figure 10.25.   M–T graph of large deep-focus earthquakes and large shallow earthquakes in Region B. Symbols are the same as in Fig. 10.24. Except for one deep focus earthquake marked with a star, all these deep-focus earthquakes occurred prior to great shallow earthquakes.

circles are all deep-focus earthquakes of M $\geq$ 7.0 and focal depth h $\geq$ 300 km, open double circles are all great shallow earthquakes of M $\geq$ 8.0 and a open circle is the 1993 Kushiro-oki earthquake (M 7.5, h $=$ 100 km). The deep-focus earthquakes preceding the large shallow earthquakes are connected by a solid line. Only one deep-focus earthquake (1939, M 7.1), marked with a star, occurred independently.

Figure 10.25 shows the M–T graph of large deep-focus earthquakes and large shallow earthquakes in Region B. Symbols are the same as Fig. 10.17. Except for one deep-focus earthquake marked with a star, all these deep-focus earthquakes occurred prior to great shallow earthquakes. As already mentioned, large shallow earthquakes in the subduction zone are not always preceded by marked deep-focus foreshocks. However, since great shallow earthquakes near the connection region of two deep sea trenches are preceded with high probability by deep-focus foreshocks, this result may be useful for intermediate-term earthquake prediction. It is noted that all great earthquakes of M 8.0 and above (three earthquakes) in region B were preceded by deep-focus earthquakes of M 7 class. In the section 10.3.c, data in the Earthquake Catalogue by Japan Meteorological Agency are used.

REFERENCES

Ishikawa, Y. (2004). SEIS-PC for Windows XP.

JMA. (1993). The 1993 Kushiro-oki earthquake (January 15, M7.8). Rep. Cood. Comm. Earthquake Predict., **50**, 8–16. (in Japanese).

Kasahara, M. (2000). 1993 Kushiro-oki earthquake. In: The Course of 30 years of thee Coordinating Committee for Earthquake Prediction, 149–159.

Kanamori, H. (1977). The energy release in great earthquakes. J. Geophys. Res., **82**, 2981–2987.

Kanamori, H. (1978). Quantification of earthquakes. Nature (London), **271**, 411–414.

Kanamori, H. and D.L. Anderson. (1975). Theoretical basis of some empirical relations in seismology. Bull. Seismol. Soc. Am., **65**, 1073–1095.

Mogi, K. (1973). Relationship between shallow and deep seismicity in the western Pacific region. Tectonophysics, **17**, 1–22.

Mogi, K. (1974). Active periods in the world's chief seismic belts. Tectonophysics, **22**, 265–282.

Mogi, K. (1979). Global variation of seismic activity. Tectonophysics, **57**, T43–T50.

Mogi, K. (1981). Earthquakes – Searching for their causes. Tokyo Univ. Press, pp. 164 (in Japanese).

Mogi, K. (1988). Downward migration of seismic activity prior to some great shallow earthquakes in Japanese subduction zone – A possible intermediate-term precursor, PAGEOPH, **126**, 447–463.

Mogi, K. (1993). Some regularities of seismicity pattern: A Review, Continental Earthquakes, IASPEI Publication Series for the IDNDR, vol. 3, 31–61, Seismol. Press, Beijing.

Mogi, K. (1994). Earthquake in India – 1993 Latur earthquake (M6.3), Zisin J., **18**, 1–7. (in Japanese).

Mogi, K. (2001). Lecture of Earthquakes, Asakura-Shoten, Tokyo, pp.150 (in Japanese).

Mogi, K. (2004). Deep seismic activities preceding the three large 'shallow' earthquakes off south-east Hokkaido, Japan – the 2003 Tokachi-oki earthquake, the 1993

Kushiro-oki earthquake and the 1952 Tokachi-oki earthquake, Earth Planets Space, **56**, 353–357.

Mogi, K. (2005). Deep seismic activities preceding the 2003 Tokachi-oki earthquake of M8.0 and the 1952 Tokachi-oki earthquake of M 8.2, J. Seismol. Soc. Japan, **57**, 275–278. (in Japanese).

Raleigh, C. B., G. Bennet, H. Crag, T. Hanks, P. Molnar, A. Nur, J. Savage, C. Scholz, R. Turner and F. Wu. (1977). Prediction of the Haichen Earthquake, EOS Trans. Am. Geophys. Un., **58**, 236–272.

Richter, C. F. (1935). An instrumental magnitude scale. Bull. Seismol. Soc. Am., **25**, 1–32.

Utsu, T. (1999). Seismicity Studies: A Comprehensive Review. Univ. of Tokyo Press, 876.

Utsu, T. (2004). Destructive earthquakes in the world, CD-ROM, Utsu-kai.

Subject index

A

Active period of seismic belt, 337–342
Active volcano, 211, 214
Acoustic emission (AE), 197
AE activity, 197
AE event, 217, 218
AE in loading and unloading process,
    289–302
AE under cyclic bending, 286
  — count rate, 292, 294, 296, 297
  — energy release, 293, 298, 301
AE under cyclic compression, 277–286
Aftershock, 207, 208–210, 333
Alps-Himalaya-Sunda (seismic zone), 335,
    341–343
Amonton-Coulomb law, 307
Anisotropy of rock, 57 (see 165–181)
Arrival time, 218, 219
Asperity, 317–319, 330
Axial strain measurement, 19
Axisymmetric stress, 52

B

Beam shaped specimen, 217
Bending load, 220
Bhuj (India) earthquake, 344–346
Biaxial compression method, 310
Block movement hypothesis, 323
Bourdon gage, 17
Brittle – ductile transition, 43–48
  — A- and B-type of, 47, 48
Burst-type earthquake swarm, 266
Byerlee's law, 308

C

Characteristic earthquakes, 326–332
Characteristic seismogram, 326, 328,
    330
Circum-Pacific seismic zone, 335
Clamping end effect, 3, 6
Coefficient of
  — external friction, 38–43
  — internal friction, 38–43
Concentrated stress, 209
Confined compression test, 57, 62–66
Confined extension test, 58, 62–66
Confining pressure, 17
Conventional triaxial compression, 17

Coulomb

Coulomb equation, 38, 308
  — physical meaning of, 38–43
Coulomb law, 308
Coulomb – Mohr fracture criterion, 37, 57
Crustal deformation, 321

D

Decay curve of AE activity, 202
Decrease of the end effect, 12–14
Deformation rates, 333
Deep focus earthquake, 346–357
Degree of fracturing, 212, 213
Degree of heterogeneity, 204, 243
Degree of stress concentration, 211
Distant region (from epicenter), 344, 345
Double-shear type loading machine, 311
  — conventional method, 311, 312
  — new method, 312–315
Downward migration, 352
Ductility, 43 (see triaxial compression)
Dynamic range of AE measurement, 250

E

Earthquakes (as large AE events), 197·
Earthquake disaster, 337
Earthquake repeaters, 330
Earthquake sequence, 207, 208, 213, 214
Earthquake swarm, 207, 208, 211, 213
Effect of the intermediate principal stress, 57,
    65–186
Effect of structure of medium, 240
Effect of previous loading
  — hydrostatic pressure, 27
  — axial compression, 28
Effect of b value in AE
  — of frequency range, 250–257
  — of stress level, 259, 260
Elastic shock, 197

F

Failure criteria, 113
Fault, 4, 137–153
Fault surface, 333
Finite element method, 278, 279
Foreshock, 210, 212, 213, 236, 343, 345, 346,
    349
Four point bending, 217, 286
Fracture (in the earth's crust), 332, 333
Fracture angle, 6–14

Fracture (crack) propagation process, 217, 236
Fracture strength, 3–14 (see Chapters 2 and 3)
Fracturing time, 202, 204
Friction, 307–315

G
Geotectonic structure, 212
Grain size of marble, 233
Great earthquake, 335, 341
Griffith (and modified) theory, 57
Gutenberg-Richter equation, 239

H
Haichen (China) earthquake, 343, 344
Heterogeneous medium, 210
History of triaxial experiment, 51
Homogeneous fault, 311
Homogeneous medium, 207
Hydrophone, 263

I
Inada granite (AE), 202–204, 222
Initial motion
— of P-wave, 217
— of S-wave, 217
Inner stress source, 209
Intermediate principal stress (σ_2), 17
(see Chapter 3)
Intrusion of magma, 211
Ishimoto–Iida equation, 239, 327, 329, 333
Izu Bonin Trench, 353, 354
Izu-Oshima-Kinkai (Japan) earthquake, 343, 344
Izu earthquake swarm (1980), 263

J
Japan Trench, 353

K
Kaiser effect, 197, 278
Kármán-type triaxial test, 52
Kita-izu (Japan) earthquake, 212, 213
Koyna (dam) earthquake, 344–346
Kurile Trench, 347, 353
Kushiro-oki (Japan) earthquake, 347, 351, 352

L
Latur (India) earthquake, 344–346
Length/diameter ratio, 6–12
Load cell, 18
Loading of sinusoidal waveform, 286

M
m or b value, 239
Magma chamber, 321, 326
Magnitude-frequency relation, 239, 295, 300

Main crack, 223, 224, 236
Main rupture (fracture), 218, 222, 298
Main shock, 212, 213
Maximum principal stress (σ_1), 17 (see strength)
Mechanical barrier, 244
Mechanical structure (Fault), 333
Microseismic activity, 197
Migration, 227
Mimatsu Diagram, 328
Minimum principal stress (σ_3), 17 (see Chapters 2 and 3)
Mixture ratio, 244
Model experiment, 206, 208, 235
Mohr envelope, 35
Mogi failure criteria, 122–131
Mogi model (volcano), 321, 324, 326, 327
Mogi type true triaxial machine, 56, 69–71
— design of, 67
— test specimen of, 72–74

N
Nankai-Suruga trough, 330–332
Number of cyclic loads, 287
Nonuniform medium, 209

O
Oceanic tide and earthquakes, 273–277
Ōmori formula for aftershock frequency curve, 205

P
Pakistan earthquake, 338, 342
Paraffin block, 221
Permanent strain, 26
Pine resin, 241
Plate boundary, 339–342
Precise measurement of fracture strength, 3
precursory phenomena, 343, 346, 349
Precursory strain prior to stick-slip, 317–319
Pre-existing fault plane, 333
Pre-slip at asperities, 317–319
Pressure dependence of strength, 32
— of silicate rock, 32
— of carbonate rock, 33

Q
Quaternary volcano, 214
Quiet period of seismic belt, 339, 341, 342

R
Recovery of AE activity, 283
Regular crack pattern, 241, 249
Rhythmical change (seismicity), 341
Rock burst, 197
Rock physics, 217

S

Sakurajima (Japan) Volcano, 321–326
Sample shape, 4
    short cylinder, 3, 11
    dog-bone shape, 3, 4
    long cylinder with epoxy fillet, 3, 5
Sanriku-oki (Japan) region, 329–331
Seismic fault in Izu region, 268, 269
Shinkomatsu andesite (AE), 280
Showa-Shinzan (new mountain, Japan), 327–329
Silicon rubber for jacketing, 17
Slow slip, 333
Source location (AE), 217, 222, 280, 288, 292, 302
Source region (AE), 223, 224, 226, 227
Stick-slip, 307, 316–319, 327–333
Strain gage, 18
Strength, 6 (see fracture stress, yield stress)
Stress concentration, 3, 5, 211, 278
Stress – strain relation, 19–23 (see Chapter 3)
Subducting slab, 353–355
Sumatra earthquake, 336–339, 341–343
Surveyor ship, 263

T

Tensile crack, 225
Tertiary folded region in Japan, 212
Three patterns of AE activity, 205
Three types of earthquake sequences, 206–208
Tokachi-oki (Japan) earthquake (2003), 346–350, 352, 355, 356
Tokachi-oki (Japan) earthquake (1952), 346, 347, 349, 350, 352, 355, 356
Transducer (AE), 218
Transition probability, 202
Triaxial cell by Hojem and Cook, 55
Triaxial compression and extension, 56–66
True triaxial compression, 52–56, 66–186
True triaxial compression test
    dilatancy in, 151, 154–160, 177–179
    ductility in, 75, 81, 87, 95, 131–134
    fracture angle in, 144, 148, 149
    fracture stress (strength) in, 74, 75, 80, 81, 83, 84, 87, 89, 94, 95, 97-99, 104, 106, 107, 109–112, 158, 164, 169, 171–173, 175, 182–184
    fracture pattern in, 137, 144, 146, 147, 149, 150, 152, 153, 164, 180
    strain hardening in, 75, 134
    stress drop at fracture in, 75, 82–84, 89, 96–99, 105

stress–strain curve in, 77–79, 84–86, 90–93, 108, 120, 159, 160, 163, 168, 171, 176–179
yield stress in, 75, 103, 107, 109, 111, 113–120, 128–131
Tested rocks in the true triaxial compression
    Chichibu green schist, 166–181
    Dunham dolomite, 60, 61, 63–65, 68, 75–84, 106, 107, 113–115, 123, 126–129, 131–145
    Inada granite, 98, 108–110, 125, 126, 131, 152, 158, 159
    Izumi sandstone, 183
    Karoo dolerite, 55
    KTB amphibolite, 184, 185
    Manazuru andesite, 96, 105–107, 125, 126, 131
    Mizuho trachyte, 94, 79, 105, 124, 126, 131, 150, 155–158
    Orikabe monzonite, 98, 99, 111, 112, 126, 153
    Shirahama sandstone, 182
    Solnhofen limestone, 60, 61, 63–65, 82, 84–89, 108, 116, 117, 123, 126, 127, 129, 131, 133, 134, 145–148
    Westerly granite, 182, 184
    Yamaguchi marble, 88, 90–98, 118–120, 124, 126, 128, 130, 131, 133–136, 148, 149, 159, 160
    Yuubari shale, 183
Two type AE events, 285

U

Unzen (Japan) Volcano, 329, 330
Upward migration, 352
Usu (Japan) Volcano, 327–329

V

Velocity of P-wave, 221
Viscous magma, 327–329
Volcanic eruption, 327, 328
Volcanic pumice particles, 202
Volcanic region (in Japan), 212, 214

W

Waveform (of AE), 296

Y

Yield criterion, 128–131
Young's modulus, 23–26